6th International Conference on Inter-Disciplinary Underground Science and Technology (i-DUST 2016)

E3S Web of Conferences Volume 12 (2016)

Avignon, France
1 - 3 June 2016

Editors:

K. Coulie
G. Micolau
P. Febvre

ISBN: 978-1-5108-3361-6

Printed from e-media with permission by:

Curran Associates, Inc.
57 Morehouse Lane
Red Hook, NY 12571

Some format issues inherent in the e-media version may also appear in this print version.

Printed by Curran Associates, Inc. (2016)

For additional information, please contact EDP Sciences – Web of Conferences at the address below.

EDP Sciences – Web of Conferences
17, Avenue du Hoggar
Parc d'Activité de Courtabœuf
BP 112
F-91944 Les Ulis Cedex A
France

Phone: +33 (0) 1 69 18 75 75
Fax: +33 (0) 1 69 28 84 91

contact-edps@webofconferences.org

Additional copies of this publication are available from:

Curran Associates, Inc.
57 Morehouse Lane
Red Hook, NY 12571 USA
Phone: 845-758-0400
Fax: 845-758-2633
Email: curran@proceedings.com
Web: www.proceedings.com

WEDNESDAY JUNE 1ST 2016

08:30 REGISTRATION

09:00 CONFERENCE OPENING

SESSION A **Session Chair**
Benjamin Canuel, *LP2N, Université Bordeaux/IOGS/CNRS:UMR 5298, France*

09:30 I1 **MIGA and very long baseline atom interferometry at LSBB.....N/A**
Andrea Bertoldi [1]
[1] *LP2N, Université de Bordeaux*

10:05 P1 **Forced oscillations of the mesopause by electric field quake precursor during strong solar activity : [SQUID]2 observation of the Nepal quakes, April 25th, 2015.....N/A**
Georges Waysand[1], Elisabeth Pozzo di Borgo[2,3], Alain Cavaillou[1]
[1] *Laboratoire Souterrain Bas-Bruit, La Grande Combe, F-84000 Rustrel,* [2]*Université d'Avignon et des Pays de Vaucluse,* [3] *UMR 1114 INRA-UAPV EMMAH*

10:30 Coffee Break

10:55 P2 **GPR velocity tomography characterization of Urgonian limestone (LSBB - Rustrel).....N/A**
Andriani Monica-Vita[1], Garambois Stéphane[2], Rousset Dominique[1], Sénéchal Guy[1]
[1] *LFCR (UMR 5150) UPPA,* [2] *ISTERRE (UMR 5275) UGA*

11:20 P3 **Sprite Initiation from Microfluctuations in the Lower Ionosphere.....N/A**
Martin Fullekrug[1], Nathan smith[2], Kuang Koh[1], Zhongjian Liu[1], Andrew Mezentsev[1], Robert Watson[1], Ivan Astin[1], Stephane Gaffet[2], Adrian Evans[1] and Michael Rycroft[1]
[1] *University of Bath, Bath, UK,* [2] *Université d'Avignon, LSBB, RustreL, France.*

11:45 P4 **Overview and outlook on Muon Survey Tomography based on Micromegas detectors for Unreachable Sites Technology (MUST2).....1**
I. Lazaro Roche[1], T. Serre[1], J.B. Decitre[1], A. Bitri[2], C. Truffert[3] & S. Gaffet[1]
[1] *LSBB UMS 3538,* [2] *BRGM,* [3] *Iris instruments*

12:10 P5 **The Mont Terri rock laboratory - 20 years of experience in a shale formation, St-Ursanne, Switzerland.....N/A**
David Jaeggi[1], Paul Bossart[1], Christophe Nussbaum[1]
[1]*Swisstopo, Rue de la gare 63, 2882 St-Ursanne, Switzerland*

12:35 Lunch

SESSION B **Session Chair**
Tony Monfret, *Géoazur, Valbonne, France*

14:00 I2 **Simultaneous geomagnetic monitoring with SQUIDs across the LSBB laboratory.....7**
Samuel Henry [1]
[1] *University of Oxford, UK*

14:35 P6 **New insight on the dynamic of the Fontaine de Vaucluse karsthydrosystem from Long Base Tiltmeter Data.....N/A**
Frédéric Boudin[1], Nolwenn Lesparre[2,3],Jean Chéry[2], Cédric Champollion[2], Charles Danquigny[4], Han Cheng Seat [5,6], Michel Cattoen [5,6], Laurent Longuevergne[7]
[1] *Laboratoire de géologie de l'ENS, UMR CNRS 8538, École Normale Supérieure, Paris, France,* [2] *Observatoire de recherche méditerranéen de l'environnement, Géosciences Montpellier, Université Montpellier Sciences et Techniques, Montpellier, France,* [3] *ArGEnCo Department,*

Applied Geophysics, University of Liege, Liege, Belgium, [4] UAPV, UMR1114 EMMAH, Avignon, France, [5] CNRS, LAAS, Toulouse, France, [6] Univ de Toulouse, INP, LAAS, Toulouse, France, [7] Géosciences Rennes, UMR CNRS 6118, Université de Rennes 1, Rennes, France.

END OF DAY 1

THURSDAY, JUNE 2 2016

SESSION D **Session Chair**
Ana Fernandes, *Centro de Ciências e Tecnologias Nucleares, Instituto Superior Técnico, Universidade de Lisboa, Portuga*

08:30 I3 **Life in Deep Sub Surface.....25**
Gaël Erauso & Grégoire Galès [1, 2]
[1] MIO, Institut Méditerranéen d'Océanologie, Aix-Marseille Université UM 110, Marseille, France, [2] Aix-Marseille Université, CEREGE, Marseille,

09:05 P14 **Lightning processes occuring inside thunderclouds and their observation with submicrosecond time resolution.....N/A**
Ivana Kolmasova [1, 2], Ondrej Santolik [1, 2], Ludek Uhlir [1], Radek Lan [1]
[1] Institute of Atmospheric Physics, Prague, Czech Republic, [2] Charles University, Faculty of Mathematics and Physics, Prague, Czech Republic

09:30 P15 **R&D of a SIMPLE Bubble Chamber for Dark Matter Searches.....33**
M. Felizardo [1], TA Girard [1], A. R. Ramos [1], I. Lazaro [2], A.C. Fernandes [1], J.G. Marques [1], A. Kling [1] and T. Morlat [1]
[1] C2TN, Instituto Superior Técnico, Universidade de Lisboa, Portugal, [2] LSBB UMS 3538, Rustrel, France

09:55 P16 **High Quality low-frequency free oscillation strain observations at the black forest underground observatory (BFO), Germany.....N/A**
Walter Zürn and Rudolf Widmer-Schnidrig [1]
[1] Black Forest Observatory

10:20 **Coffee Break**

SESSION E **Session Chair**
Guy Dumont, *The University of British Columbia, Canada*

10:45 P17 **Microbiology of shallow subsurface aquifer and carbonate rocks studied by culture-dependent and culture-independent methods: example of the LSBB, Provence, France.....39**
Grégoire Galès [1,2], Gaël Erauso [2], Daniel Boyer [3] & Gilles Micolau [4]
[1] Aix-Marseille Université, CEREGE, Marseille, France, [2] Aix-Marseille Université, Université du Sud Toulon-Var, CNRS/INSU, IRD, MIO, UM 110, Marseille, France, [3] LSBB UMS 3538, Rustrel, France, [4] UMR EMMAH, INRA/Université d'Avignon et des Pays de Vaucluse, Avignon

11:10 P18 **Particle transport and diffusion in a micro-pump.....N/A**
Mounia makhoul, Philippe Beltrame, Maminirina joelson [1]
[1] UMR 1114 University of Avignon (UAPV), INRA, Avignon, France

11:35 P19 **Measurement of lightning related electromagnetic signals in a wide frequency range.....N/A**
Ondrej Santolik [1,2], Ivana Kolmasova [1,2],Ludek Uhlir [1],Radek Lan [1]
[1] Institute of Atmospheric Physics, Prague, Czech Republic, [2] Charles University, Faculty of Mathematics and Physics, Prague, Czech Republic

12:00 P20 **Is groundwater a potential reservoir for the human and plant bacterial pathogens, bacillus cereus and pseudomonas syringae ?.....N/A**
Ludivine Rousset [1,2], Odile Berge [2], Marina Gillon [3], Vincent Marc [3], Anne-Laure Cognard-Plancq [3], Thierry Clavel [1], Catherine Duport [1]

[1] *SQPOV, UMR0408, Avignon Université, INRA, Avignon, France,* [2] *MISTRAL, UR0407, INRA, Avignon, France,* [3] *EMMAH, UMR1114, Avignon Université, INRA, Avignon, France*

12:25 **Lunch**

SESSION F **Session Chair**
Matthew Yedlin, *University of British Columbia, Canada*

14:05 I4 **LSBB, a unique underground laboratory for carbonate reservoir caracterisation - Example of the HPMS-Ca Project: Predicting Hydraulic, Petrophysical, Mechanical and Seismic Properties of Carbonates.....N/A**

Yves Guglielmi [1], Jean Borgomano [1]

[1] *Aix-Marseille Université, CEREGE, Marseille, France*

14:40 P21 **Ultra-low Noise EEG at LSBB: New Results.....44**

N. Hamzei [1], J. Steeves [1], M. Yedlin [1], GA Dumont [1]

[1] *The university of British Columbia, Canada*

15:05 P22 **Noise Modeling in a Signal Conditioning Circuit for Low Power Audio Application Using Resistive Sensor.....53**

Stéphane Meillère [1], Eric Savary [1], Wenceslas Rahajandraibe [1], Hervé Barthélémy [1]
[1] *IM2NP, Aix-Marseille Université, Marseille*

15:30 P25 **On the investigation of voltage controlled oscillator phase noise for IoT applications.....62**

Fayrouz Haddad [1], Wenceslas Rahajandraibe [1]

[1] *IM2NP, Aix-Marseille Université, Marseille*

15:55 **Coffee Break**

SESSION G **Session Chair**
Samuel Henry, *University of Oxford, UK*

16:20 P24 **Statistical investigations of the Nickel film morphology by Atomic Force Microscopy.....69**

Rosine Coq-Germanicus [1], Philippe Leclère [2], Eric Hug [3], Philippe Descamps [1]

[1] *Normandie Université ENSICAEN /CRISMAT /UMR 6508, Caen, France,* [2] *Service de Chimie des Matériaux Nouveaux, CIRMAP Université de Mons, Belgium,* [3] *CRISMAT, Normandie Université UMR CNRS 6508, Caen, France*

16:45 P23 **Polarisation Properties of [SQUID]2 Horizontal Components at the LSBB (Laboratoire souterrain a bas bruit).....75**

Christian Kwisanga [1]

[1] *University of Rwanda, College of Science and technology, Rwanda*

17:10 P26 **Progress on Ultra-wideband Multi-Antenna Radar Imaging for MIGA.....82**

M. Yedlin [1], J.Y. Dauvigniac [2], N. Fortino [2], Ch. Pichot [2], R. Cova [3], D. Henley [3], K. Hall [3]

[1] *University of British Columbia,* [2] *Laboratoire d'Electronique, Antennes et Télécommunications,* [3] *University of Calgary*

17:35 P27 **Some technical requirements for insuring valuable measurements toward a 3 Component and 3 Dimension, 3C-3D, seismic array processing.....N/A**

Olivier Sèbe [1], Claire Labonne [1,2], Stéphane Gaffet [2,3], Jean-Baptiste Decitre [3], Daniel Boyer [3], Jean-Marc Koenig [1], François Schindelé [1]

[1] *CEA, DAM, DIF, Arpajon, France,* [2] *Univ. Nice Sophia Antipolis, CNRS, IRD, Observatoire de la côte d'Azur, Géoazur UMR 7329, Valbonne, France,* [3] *LSBB, UMS 3538, Rustrel, France*

20:00 **Gala dinner**

FRIDAY, JUNE 3 2016

SESSION H **Session Chair**
Guy Sénéchal, *LFCR (UMR 5150) UPPA*

08:35 P28 **Low-background gamma spectrometry at the underground laboratory of Ferrières, French Pyrénées.....N/A**
Marc Souhaut [1], Pieter Van Beek [1], Simon Bejannin [1], Emilie Le Roy [1]
[1] Université Toulouse III Paul Sabatier, Observatoire Midi Pyrénées - LAFARA, LEGOS

09:00 P29 **Full-waveform Inversion of GPR data acquired between boreholes in Rustrel carbonates.....88**
H. Pinard [1], S. Garambois [1], L. Métivier [1], M. Dietrich [1], G. Sénéchal [2], D. Rousset [2]
[1] STerre, Université Grenoble Alpes & CNRS (UMR 5275), Grenoble, France , [2] LFCR, Université de Pau et des Pays de l'Adour, Pau, France

09:25 P30 **Atmospheric Radiation Environment Analyses Based-on CCD Camera at Various Mountain Altitude and Underground sites.....95**
Pierre Li Cavoli[1,2], Guillaume Hubert [2], José Busto [1]
[1] the French Aerospace Laboratory (ONERA), Toulouse, France, [2] The Center for Particle Physics of Marseille, CNRS/IN2P3, Marseille, France

09:50 P31 **In-Situ Exploration of Hydromechanical properties of Porous and Fractured Carbonates Rocks at Metric Scale.....N/A**
S. Séguy [1], V. Guinot [1], F. Cappa [2], Y. Guglielmi [3], H. Jourde [1], J.P. Rolando [4], G. Massonnat [4]
[1] HydroSciences Montpellier, UMR 5569 CNRS-IRD-UM1-UM2, Université Montpellier 2, France, [2] Université de Nice Sophia-Antipolis, Observatoire de la Côte d'Azur, GeoAzur, Sophia-Antipolis, France, [3] Aix-Marseille Université, CEREGE, IRD, CNRS, Marseille, France, [4] Total SA, Avenue Larribau, Pau, France

10:15 **Coffee Break**

SESSION I **Session Chair**
Pascal Febvre, *IMEP-LAHC, CNRS UMR5130, University Savoie Mont Blanc, France*

10:40 P32 **Toward an innovative stochastic modeling of electric charges loss through dielectric 102**
G. Micolau [1], J. Postel-Pellerin [2], P. Chiquet [2], M. Joelson [1], C. Abbas [1], D. Boyer [2], C. Ginoux [1]
[1] UMR EMMAH, University of Avignon, [2] IM2NP University of Marseille, [3] LSBB, Rustrel

11:05 P33 **Detection of alpha particle contamination in ULA-grade integrated circuits.....111**
A.C. Fernandes [1], T.A. Morlat [1], M. Felizardo [1], R.C. Martins [2], A. Kling [1], J.G. Marques [1], T.A. Girard [1,3]
[1] Centro de Ciências e Tecnologias Nucleares, Instituto Superior Técnico, Universidade de Lisboa. Campus Tecnológico e Nuclear, Bobadela, Portugal [2] Instituto de Telecomunicações, Instituto Superior Técnico, Universidade de Lisboa. Lisboa, Portugal, [3] Departamento de Física, Faculdade de Ciências, Universidade de Lisboa. Campo Grande, Lisboa, Portugal.

11:30 P34 **Geochemical and hydrodynamic characteristics of the karstic system of Jbal Zaghouan (Northeastern-Tunisia).....N/A**
RandaDziria [1], NajibaChkira [2], Christophe Emblanch [3], Kamel Zouaria [4], Abdelrahmen Gallalid [5]
[1] Laboratoire de radio-analyses et environnements de l'école d'ingénieurs de Sfax, Sfax, Tunisie, [2] Département de géographie, laboratoire de radio-analyses et environnement de l'école d'ingénieurs de Sfax, Faculté des lettres et sciences humaines de Sfax, Tunisie, [3] UMR EMMAH, UFRip STS, Université d'Avignon, Avignon, France, [4] Commissariat Régional de développement Agricole de Zaghouan, Zaghouan

11:55 P35 Relationship between diagenetic history and fracturation in carbonates.....N/A

J. Cochard[1], Y. Guglielmi[1], P. Léonide[1], J. Lamarche[1], J.P. Rolando[2], G. Massonnat[2]

[1] Aix-Marseille Université, CEREGE, IRD, CNRS, Marseille, France, [2]Total SA, Pau, France

12:20 P36 Analysis of different magnetic field noise environments for preparing a long-base highly-sensitive magnetic interferometer.....N/A

Pascal Febvre [1], Ali Bozbey [2], Deniz Balaban [2], Sasan Razmkhah [2], Cengiz Çelik [2], Elisabeth Pozzo di Borgo [4,5], E. F. Saunderson [6], D. J. Gouws [5] and C.J. Fourie [7]

[1] IMEP-LAHC, CNRS UMR5130, University Savoie Mont Blanc, Le Bourget du Lac, France, [2] Department of Electrical and Electronics Eng., TOBB University of Economy and Technology, Ankara, Turkey, [3] LSBB, UMS 3538 UNS/UAPV/CNRS/AMU/OCA, Rustrel, France, [4] UMR1114, UAPV-INRA, EMMAH, Avignon, France, 5 UMR1114, INRA-UAPV, EMMAH, Avignon, France, [6] South African National Space Agency, Space Science Directorate, Hermanus, South Africa, [7] Stellenbosch University, Stellenbosch, South Africa

12:45 CLOSING REMARKS

13:00 Lunch

Additional Papers

E3S Web of Conferences
Open access conference proceedings
in Environment, Energy & Earth Sciences
A flexible, fast and efficient
publication service,
for an open access recording
of your event
and a wide dissemination of
your proceedings
www.e3s-conferences.org
edp open Open Access Publishing from EDP Sciences

Aims & scope

E3S Web of Conferences is an open-access journal devoted to the publishing of conference proceedings dealing with Environment, Energy and Earth Sciences.

Coverage includes technological and scientific aspects as well as social and economic matters. Major disciplines include: soil science, hydrology, oceanography, climatology, geology, geography, energy engineering (production, distribution and storage), renewable energy, sustainable development, natural resources management, etc.

The journal aims to make conference proceedings visible, citable and widely usable by the scientific community. It offers outstanding services for the publication and dissemination of contents in order to ensure long-term archiving and maximum exposure of scientific events.

Why publish with E3S Web of Conferences?

- **Open access:** conference papers available worldwide.
- **Fast publication:** less than 3 months.
- **Total flexibility:** no page restrictions, free unlimited use of colour in the online version, any kind of scientific document welcomed (traditional articles, posters, abstracts and slideshows).
- **High visibility:** all documents indexed in main databases and easily found online.
- **Fair author rights:** CC-BY Creative Commons license.
- **Range of additional publishing options:** printed proceedings, CD-ROMs, copy-editing, etc.

Contact: Isaline Augusto
e3s@webofconferences.org

ISSN (electronic edition): 2267-1242

www.e3s-conferences.org

edp open Open Access Publishing from EDP Sciences

Committees and List of reviewers

Organizing Committee

Daniel	Boyer	LSBB – UMS 3538 CNRS, France
Alain	Cavaillou	LSBB – UMS 3538 CNRS, France
Karine	Coulié	IM2NP UMR 7334, Aix-Marseille Université, France
Pascal	Febvre	IMEP-LAHC – CNRS UMR5130, Université Savoie Mont Blanc, France
Gilles	Micolau	EMMAH, Université d'Avignon et des Pays du Vaucluse, France

Scientific Committee

Ali	Bozbey	TOBB University of Economics and Technology, Turkey
José	Busto	CPPM UMR 7346, Aix-Marseille Université, France
Coenrad	Fourie	Stellenbosch University, South Africa
Martin	Fullekrug	University of Bath, United Kingdom
Grégoire	Galès	CEREGE, Aix-Marseille Université, France
Samuel	Henry	University of Oxford, United Kingdom
Maminirina	Joelson	EMMAH, Université d'Avignon et des Pays du Vaucluse, France
Laurent	Longuevergne	Géosciences Rennes, Université de Rennes, France
Gérard	Massonnat	Total
Laurent	Ottaviani	IM2NP UMR 7334, Aix-Marseille Université, France
Jérémy	Postel-Pellerin	IM2NP UMR 7334, Aix-Marseille Université, France
Wenceslas	Rahajandraibe	IM2NP UMR 7334, Aix-Marseille Université, France
Jean-Paul	Rolando	Total
Frédéric	Wrobel	IES UMR5214, Université de Montpellier, France

List of Reviewers

Odile	Berge	INRA, UR 0407 PV Pathologie Végétale, Avignon, France
Benjamin	Canuel	Institut d'Optique d'Aquitaine, LP2N UMR 5298, Talence, France
Rosine	Coq-Germanicus	CRISMAT-LaMIPS Laboratory, UMR 6508, Normandy University UCBN, France

Luigi	Dilillo	LIRMM UMR 5506, Montpellier, France
Guy	Dumont	Department of Electrical and Computer Engineering, The University of British Columbia, Vancouver, Canada
Dominique	Fouchez	Aix Marseille Université, CNRS/IN2P3, CPPM UMR 7346, 13288, Marseille, France
Ivana	Kolmasova	Institute of Atmospheric Physics, Prague, Czech Republic
Gaelle	Lefeuve-Mesgouez	EMMAH, Université d'Avignon et des Pays du Vaucluse, France
Pascal	Masson	LEAT UMR 6071, Université de Nice, France
Stéphane	Meillère	IM2NP UMR 7334, Aix-Marseille Université, France
Laurent	Nony	IM2NP UMR 7334, Aix-Marseille Université, France
Laurent	Ottaviani	IM2NP UMR 7334, Aix-Marseille Université, France
Benoît	Paternoster	Total
Wenceslas	Rahajandraibe	IM2NP UMR 7334, Aix-Marseille Université, France
Elda	Saunderson	South African National Space Agency

i-DUST 2016 sponsors

UNIVERSITÉ D'AVIGNON

UNIVERSITÉ D'AVIGNON ET DES PAYS DE VAUCLUSE

IRSN

INSTITUT DE RADIOPROTECTION ET DE SÛRETÉ NUCLÉAIRE

Aix*Marseille université

Université Nice Sophia Antipolis

Foreword

The 6th edition of the inter-**D**isciplinary **U**nderground **S**cience & **T**echnology (i-DUST 2016) international conference took place in the wonderful premises of the University d'Avignon et du Pays de Vaucluse in the centre of Avignon on June 1–3, 2016, following the annual user's day of LSBB that was hold on May 31st. i-DUST 2016 welcomed 62 participants from Canada, Czech Republic, France, Germany, Portugal, Rwanda and United Kingdom. Several topics related to underground science were discussed in five sessions: "Geological reservoirs", "Couplings", "Particles detection and measurement", "Microelectronics materials, devices and systems", "Biology", plus an additionnal session for "open fields". Out of 39 presentations including 4 invited talks, 19 publications were reviewed and accepted for publication in **E3S Web of Conferences**. They are all accessible in open access from this page. For this i-DUST edition, all talks from the different topics were mixed within each session to facilitate the interactions associated to inter-disciplinary science, which was much appreciated by attendees. Next i-DUST 2018 conference is planned in Provence on spring 2018, either in Avignon or in another town around LSBB, maybe Cadarache or Aix-en-Provence. We are looking forward for your contribution and attendance.

edp sciences

E3S Web of Conferences

Statement of Peer review

In submitting conference proceedings to E3S Web of Conferences, I certify to the Publisher that I adhere to the **Policy on Publishing Integrity** of the journal in order to safeguard good scientific practice in publishing.

1. All articles have been subjected to peer review administered by the proceedings editors.
2. Reviews have been conducted by expert referees, who have been requested to provide unbiased and constructive comments aimed, whenever possible, at improving the work.
3. Proceedings editors have taken all reasonable steps to ensure the quality of the materials they publish and their decision to accept or reject a paper for publication has been based only on the merits of the work and the relevance to the journal.

Title, date and place of the conference

i-DUST 2016 (inter-Disciplinary Underground Science & Technology)

Avignon (France)

June 1-3, 2016

Proceedings editor(s):

Karine Coulié - Aix-Marseille Université

Gilles Micolau - Université d'Avignon et des Pays du Vaucluse

Pascal Febvre - Université Savoie Mont Blanc

Date and editor's signature

November 7 2016

Karine Coulié Gilles Micolau Pascal Febvre

DOI: 10.1051/e3sconf/20161203001

Overview and outlook on muon survey tomography based on micromegas detectors for unreachable sites technology

I. Lázaro Roche[1,2,3,a], T. Serre[1], J.B. Decitre[2], A. Bitri[3], C. Truffert[1], and S. Gaffet[2]

[1] Iris Instruments, 1 avenue buffon, BP. 16007, 45060 Orléans Cedex 2, France
[2] LSBB UMS3538, La grande combe, 84400 Rustrel, France
[3] BRGM, 3 avenue Claude-Guillemin, BP. 36009, 45060 Orléans Cedex 2, France

Abstract. The present document describes the functioning principles of the Muon Survey Tomography based on Micromegas detectors for Unreachable Sites Technology and its distinguishing features from other Micromegas-like detectors. Additionally, it addresses the challenges found while operating the first generation and the resulting improvements. Currently, the project Temporal Tomography of the Densitometry by the Measurement of Muons is focused on obtaining a reliable pulse from the micromesh, associated to the passing of a muon, in order to trigger the acquisition and operate in standalone mode. An outlook of the future steps of the project is provided as well.

1. Introduction

Muons are charged particles originated naturally all around us. Primary cosmic rays from deep space hit the upper atmosphere and produce secondary particles, including muons. Muons are heavy, (207 times more massive than the e^-), fast (99.97% of the speed of light) and they lack of strong interaction. These characteristics allow them to propagate up to several hundreds of meters underground, with an absorption proportional to the amount of matter they travel trough. Muon tomography, or muography, leans on the drop of the muon population as depth goes on to calculate the density of the matter crossed [1, 2].

The Temporal Tomography of the Densitometry by the Measurement of Muons collaboration (T2DM2) has developed a new tool to measure the muon's flux and trajectories in order to image dynamically the densitometry of the medium. The MUST^2 camera is born from the inclusion of a Micromegas detector [3, 4] within a Time Projection Chamber. This combination presents interesting distinctive features compared with the regular Micromegas detectors, such as the capability to reconstruct the particle's trajectory with only one detection plane. This reduces substantially the volume and weight of the camera, specially compared to other muon detection systems, such as scintillators.

The technology validation for this new tool is being carried out by the Low Noise Inter Disciplinary Underground Science and Technology URL (LSBB) with the prototypes developed between 2012 and 2015.

[a] Corresponding author: ignacio.lazaro@lsbb.eu

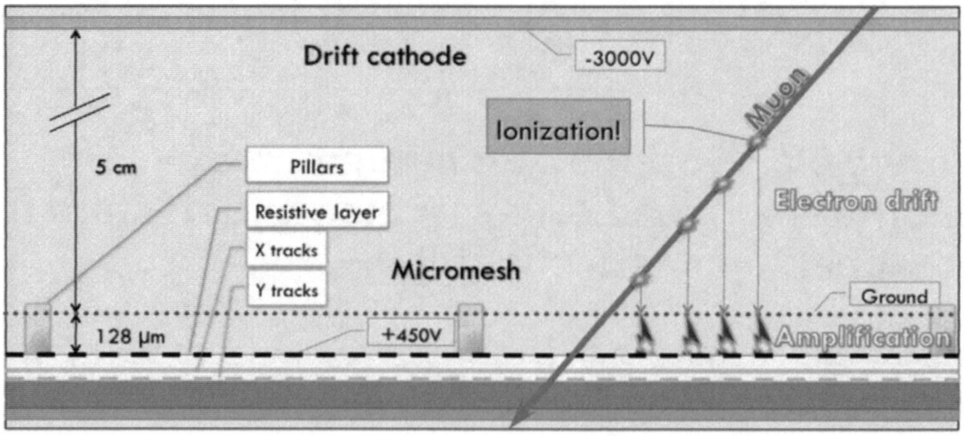

Figure 1. Schematic cross-section of the detector with its characteristic elements and functioning principle.

2. MUST² detector

The Muon Survey Tomography based on Micromegas detectors for Unreachable Sites Technology (MUST^2) is a micro-pattern gaseous detector with two separated volumes: a conversion & drift region, acting as Time Projection Chamber (TPC), and an amplification gap [5]. In the lower part of this gap lays the resistive layer, which serves as protection against sparks and reduces the output current that may damage the frontend electronics. Underneath the resistive layer, there are two superimposed layers with the readout tracks of X and Y axis respectively. The two planes are twisted 90° with respect to each other to obtain a 2D measurement.

As seen in Fig. 1, when the ionizing particle enters in the conversion chamber, it ionizes the gas and the electrons generated will drift orthogonally towards the micromesh under the influence of the electric field in this region. Once the electron goes trough the micromesh, it enters in the amplification gap with a much stronger electric field, where undergoes acceleration that creates an avalanche effect that amplifies the charge. This charge, when collected by the resistive layer, will induce with its displacement a signal by capacitive coupling in both the micromesh and the readout tracks. The first one is used to command the acquisition and the second one for the image reconstruction.

By knowing the drift speed of the electron trough the gas and the arrival times of the ionizations, we are able to infer the zenithal angle of the trajectory.

Both the choice of gas and drift voltage for this topology are motivated by the need of a fast drift of the primary electrons in order to fit the whole signal in the 675 ns time window of the electronics, restriction imposed by the buffer size of the hybrid cards [6]. It is possible to determinate precisely the drift velocity of electrons in gas mixtures, for instance with the Magboltz code [7], to establish the operational parameter. The selected blend is composed of argon, CF4 and isobutene (88%, 10% and 2% respectively), which has a peak of drift speed at 600 V/cm of more than 10 cm/μsec.

The gain of the detector is variable and is determined by the choice of the gas and the electric field in the amplification gap (controlled by the voltage of the resistive layer). The gain should be as high as possible but never exceed the Raether limit [8]. Otherwise the avalanche process becomes unstable and leads to the creation of a thin plasma filament that may damage irretrievably the resistive tracks. Typically the detector operates with a resistive voltage near the 420 V and it has a gain in the order of 10.000 [9].

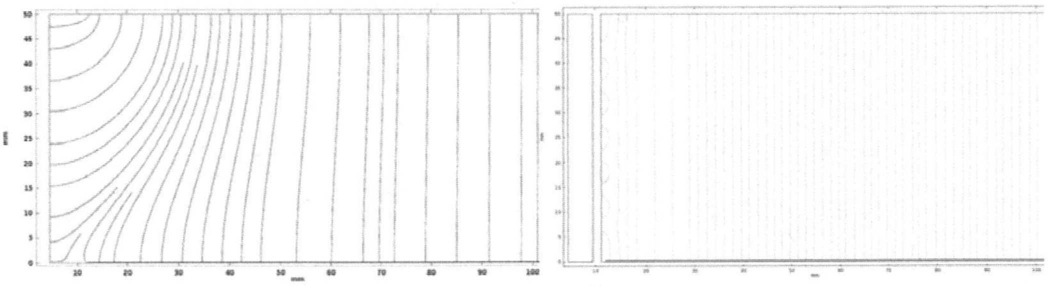

Figure 2. Field lines of the drift volume with (right) and without (left) the homogenizer device obtained with COMSOL Multiphysics®.

The ratio between the two electric fields should be evaluated as well because it also influences the capability of the electrons to travel from the drift region to the amplification gap without being stopped (mesh electron transparency [10]), which creates a loss of information.

3. Challenges

The MUST^2 presents some distinguishing factors compared with other Micromegas detectors. It has a drift space of 5 cm, which is about ten times the height of the most extended configuration for this kind of detector. On the one hand, this feature allows the reconstruction of the particle track with only one detection plane, with a resolution of 1 and 2 degrees in zenithal and azimuthal angles respectively. On the other hand, as shown in Fig. 2, the big distance from the cathode to the micromesh creates a non-homogeneous electric field all along the perimeter of the detector in the first 5 cm from the external frame.

To remove the possible artifacts induced by the local heterogeneities, an electric field homogenizer has been designed, based on COMSOL Multiphysics ® simulations, and placed inside the TPC. This element has the double function to create a field gradation in the perimeter of the detector and, at the same time stimulates the gas distribution.

A new type of resistive layer has been recently developed by the Kobe University. The preliminary tests [11] with the latest Diamond-Like Carbon deposition resistive layers (DLC), have proven its robustness and good performance, creating a reliable alternative to the classical Carbon-Loaded Kapton layers (CLK). Some of their interesting features are the bigger charge collected by the hybrid cards, due to the bigger capacitance per unit of area of this layer, and the lower charge dispersion due to a smaller surface resistivity. They also show similar position resolution.

Electric field simulations of the amplification gap for both layers show that the DLC doesn't present the distortions in the amplification gap (see Fig. 3) due to the relief of the CLK.

A prototype of bulk-Micromegas with a DLC resistive foil has been already commissioned to the CERN PCB workshop and it's expected before the end of 2016.

4. Self trigger

Obtaining a trustworthy trigger signal to control the acquisition is one of the key settings of the detector. The device is able to acquire data using an external trigger, for instance created

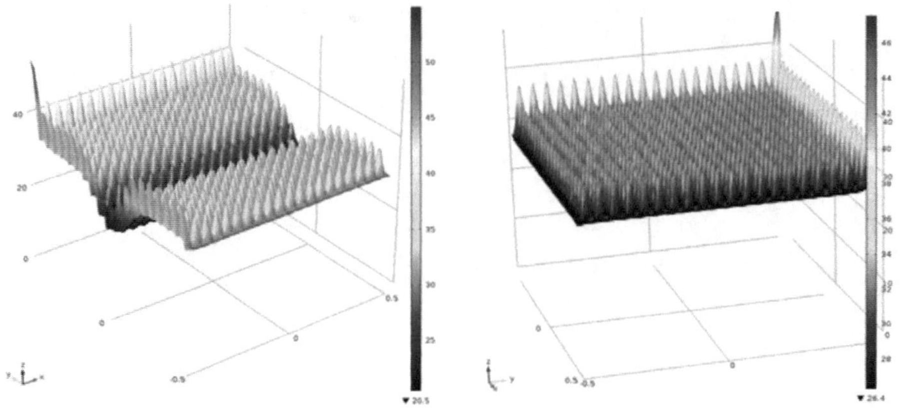

Figure 3. Simulated voltage in the micromesh plane with the CLK (left) and DLC (right) resistive layers obtained with COMSOL Multiphysics ®. X, Y, Z axis in cm and values in V.

with scintillators, but it has been conceived to be able to produce an internal trigger in order to operate in stand-alone mode. This reduces the number of peripherals and hence, the cost and volume of the detector.

The signal collected from the Micromesh is amplified and sent to a discriminator, where it is possible to set manually the trigger threshold. When the signal exceeds the fixed level, it creates a NIM pulse [N] that controls the FPGA and records the information of the readout tracks.

The Fig. 4 shows the preliminary results of the signal obtained from the mesh by measuring cosmic rays.

The $MUST^2$ was placed between two plastic scintillators and the Fig. 4 shows the coincidence events between the three detectors in order to validate the signal obtained from the micromesh. The gas blend used in the experience is Ar/CO_2 (93:7), the drift electric field is set at 600 V/cm and the voltage applied to the resistive layer ranges between 500 and 560 V (560V in the case of Fig. 4) to assess its effect in the gain. The micromesh signal is preamplified with an Ortec 142 C and amplified with an Ortec 471.

While operating with only the resistive potential (which means no muons converted into electrons), a peak of noise appears at ~204 mV. When the drift is activated, the events over 210 mV start to appear and create for every voltage a similar shape, but with an amplitude proportional to the gain of the detector. This behavior is consistent with the functioning principle of the detector and the signal so generated may be used to trigger the acquisition with a delay smaller than the buffer time of the electronics.

Nevertheless, further analysis with different gases and experimental parameters needs to be done before we can fully validate the self-trigger feature.

5. Summary and conclusion

The current stage of the project consists of incorporating the improvements resulting of our experience with the physics simulation [12] and the first generation to overcome some limitations:

- A slight redesign of the mechanical components to minimize gas leaks.
- The replacement of the *bulk* Micromegas for a *floating-mesh* one. This will increase the ease of servicing in case of contamination of the detector, and will extend the lifetime of the detector.

DOI: 10.1051/e3sconf/20161203001

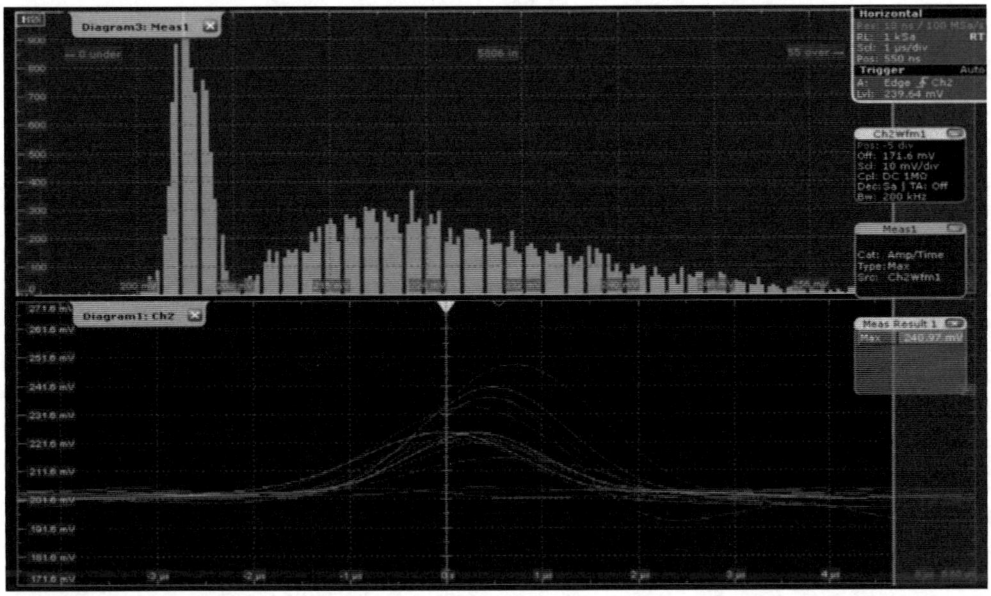

Figure 4. Oscilloscope's print-screen with the characteristic signal shape (bottom) and the histogram of signal amplitudes (top).

- The use of optic fiber equipment. It will notably increase the distance required between the detector and the data acquisition station. Hence reducing duplicity of materials in spread networks.
- The use of custom-made electronics that fits more precisely our needs.

Muon tomography, based on the absorption or attenuation of cosmic muons, is a mature technique still opened to innovation and the development of promising applications. The high temporal and spatial resolution of the new MUST^2 camera allows to get more detailed images within less time. Thanks to its compact size and lightness, it is possible to deploy the device in places unreachable for other muon measurement equipment. It can be used in the framework of natural risk management and for Earth science and public works applications. This technology complements the existing ones (seismic, gravimetry, electric…). Once the camera prototype is fully operational, the collaboration will be focused on demonstrating this complementarity in real study cases.

The authors thank the priceless assistance of the CERN's PCB workshop and the RD51 collaboration for the development of micro-pattern gas detectors technologies for allowing us to work in their facilities with their equipment. Special thanks to R. de Oliveira, H. Muller and E. Oliveri for their commitment and dedication. This research was performed within the framework of a CIFRE grant (ANRT contract #2015/0325) and supported by the European Fund for Regional Development (FEDER).

References

[1] E.P. George, *Cosmic rays measure overburden of tunnel*, Commonw. Eng. 455–457 (1955).

[2] L. Alvarez, et al., *Search for hidden chambers in the pyramids*, Science **167**, 832–839 (1970).

[3] Y. Giomataris, Ph. Rebourgeard, J.P. Robert, G. Charpak, *A high-granularity position-sensitive gaseous detector for high particle-flux environments*, Nucl. Instr. Meth. Phys. Res A **376**, 29–35 (1996).

[4] I. Giomataris, R. De Oliveira, S. Andriamonje, S. Aune, G. Charpak, et al., *Micromegas in a bulk*, Nucl. Instr. Meth. A **560**, 405–408 (2006).

[5] I. Lázaro et al., *Muon telescope based on Micromegas detectors: from design to data acquisition, Proceedings of i-DUST conference* (2014).

[6] J.-D. Berst et al., *Recent developments and results on APV (DMILL) circuits for silicon and MSGC detectors,* Proceedings of the Fifth Workshop on Electronics for LHC Experiments, CERN 99-09, CERN/LHCC/99-33, 108–112 (1999).

[7] S.F. Biagi, *Monte Carlo simulation of electron drift and diffusion in counting gases under the influence of electric and magnetic fields,* Nucl. Instr. and Meth. A **421**, 234–240 (1999).

[8] T. Francke, *Innovative Applications and Developments of Micro-Pattern Gaseous Detectors*, ISBN 9781466660151 (2014).

[9] C. Adloff, M. Chefdeville, A. Espargiliere, R. Gaglione, *Environmental study of a Micromegas detector*, HAL Id: in2p3-00413881, pp.19 (2009).

[10] K. Nikolopoulos, P. Bhattachary, V. Chernyatin, R. Veenhof, *Electron transparency of a Micromegas mesh*. J. Instrum **6** (2011).

[11] P. Colas et al., *Test of a new anode resistive coating for a Micromegas TPC*, RD51 collaboration meeting (2015).

[12] U.S. NIM committee, *Standard NIM Instrumentation System*, DOE/ER-0457T (1990)

[13] F. Hivert et al., *Muography sensitivity to hydrogeological rock density perturbation: roles of the absorption and scattering on the muon flux measurement reliability.* Near Surf. Geophysics (accepted).

 DOI: 10.1051/e3sconf/20161202003

Simultaneous geomagnetic monitoring with multiple SQUIDs and fluxgate sensors across underground laboratories

S. Henry[1], E. Pozzo di Borgo[2], C. Danquigny[2], and B. Abi[1]

[1] University of Oxford, Department of Physics, Denys Wilkinson Building, Keble Road, Oxford, OX1 3RH, UK
[2] UAPV, UMR1114 EMMAH, INRA, 84914 Avignon, France

Abstract. For two periods of several weeks duration in spring and autumn 2014, we monitored fluctuations in the geomagnetic field using 3-axis SQUID magnetometers at three locations within the LSBB tunnel network. Measurements at the water flow point C, and the GAS gallery supplemented the measurements by the permanent [SQUID]^2 system in the Capsule. The frequency spectra of the magnetic fluctuations varied considerably between the three locations. These measurements have provided a unique data set to investigate the potential of underground SQUIDs to monitor the gradient of geomagnetic signals and thus investigate magnetotelluric effects and possible seasonal effects due to changes in the water content of the surrounding rock. We also compared the measurements of a SQUID and fluxgate system, and report on fluxgate measurements at the Boulby Underground Laboratory.

1. Introduction

The sources of time-varying geomagnetic fields include electrical currents in the ionosphere and telluric currents induced in the ground. Groundwater will influence the magnetic field as the water content of rock modifies its electrical properties. The flow of water through rock may also generate a magnetic field through the electrokinetic effect. The LSBB underground laboratory is an ideal site to study these processes, as the host of the permanent 3-axis magnetometer – [SQUID]^2 (SQUID in Shielding Qualified for Ionosphere Detection); as well as a network of seismometers and groundwater flow monitors [1].

SQUIDs (Superconducting Quantum Interference Devices) are precision magnetometers using a combination of superconducting loops and Josephson Junctions to track changes in the magnetic flux through a superconducting pick-up loop at a level equivalent to $\sim 10^{-15}$ T$\cdot$ Hz$^{-1/2}$ equivalent to $\sim$0.1 pT over a typical bandwidth of interest. However they are insensitive to the absolute value of the magnetic field and the length of time for which they can track the quasi-dc field is limited by flux jumps induced by electromagnetic interference and rapidly changing magnetic fields.

The Oxford University Department of Physics has developed new hardware and software techniques to monitor magnetic field fluctuations using SQUIDs, as part of a particle physics experiment [2, 3]. In collaboration with LSBB and UAPV we conceived a project to search for the magnetic signal from groundwater flow. We carried out a long term measurement in 2012 [4, 5]. This measurement gave a null result and set a lower limit of 0.2 nT on the magnetic field generated by groundwater flow at LSBB [6].

The measurements showed that there is a significant magnetic field gradient across the underground laboratory due to telluric currents induced in the Earth's surface by the primary ionospheric signals. This is a limiting factor when searching for a local magnetic signal, but it also highlights a new opportunity. These currents are proportional to the electrical conductivity of the surrounding rock. Short term variations in the electrical conductivity are most likely caused by changes in the water content. Therefore we can in principle probe the water content by measuring the associated magnetic signals. Such magnetotelluric measurements with our SQUID technology could potentially allow higher resolution geophysics exploration.

Magnetotellurics (MT) is an electromagnetic technique for underground exploration using the natural fluctuations in the geomagnetic field as a probe signal. By monitoring the vector components of the magnetic field and the electric field on the surface, usually by simultaneous measurements in multiple locations, the impedance tensors and hence the electrical conductivity can be determined. The frequency measured, which ranges from mHz to kHz, determines the depth which can be probed [7, 8].

MT magnetic measurements are generally done using induction coils (at higher frequencies) and fluxgate probes (at lower frequencies). SQUIDs can achieve a resolution $\sim$1000 times better than fluxgates [9] over frequencies from quasi-dc to the top of the range of interest. After some preliminary studies [10], it appears they were abandoned as MT instruments. Their ability to probe low frequency signals is limited by interference induced artefacts. These can now be corrected using software techniques. With these new data analysis techniques it may now be feasible to use SQUIDs as magnetic sensors for MT measurements.

This paper reports further measurements carried out in 2014 and gives the results of a preliminary analysis. Simultaneous measurements were taken using three SQUID systems at the CAP, GAS and C points within the LSBB laboratory. In the following sections we discuss these measurements, as well as a comparison of SQUID and fluxgate measurements at the point C, and fluxgate measurements at the Boulby Underground Laboratory in the UK.

2. Variation in geomagnetic fluctuations across the LSBB complex

Frequency spectra of the geomagnetic signals recorded at the three points are shown in Fig. 1. As expected, the fluctuations are lowest in the Capsule, where the shielding of the metal walls reduce the geomagnetic field, and measures are taken to minimise the magnetic noise from power lines and other infrastructure. The CAP spectra does show some features in the range 3–30 Hz, likely to include a contribution from artefacts and perhaps the Schumann resonances. The origin of the sharp peak at 0.16 Hz is not clear.

The GAS and C spectra are distinctly different. It is not clear why the environment is different as there is a similar height of rock above each location. The fluctuations recorded at GAS show the highest level between 0.1 and 5 Hz, above this the spectrum drops down. The C spectra shows a series of peaks at the Schumann resonances at 8, 14, 21 and 28 Hz. These are much clearer than previous reports of the Schumann resonances in data recorded by the $[\mathrm{SQUID}]^2$ system [11]. It appears these resonances are attenuated by the shielding around the Capsule, but the rock above point C has minimal effect on magnetic signals at these frequencies.

To compare the amplitude of the quasi-dc fluctuations at the different locations, we used a fitting algorithm to rescale the time domain signals recorded by the $[\mathrm{SQUID}]^2$ system to give the best fit to one-hour segments of data recorded at the other stations, using the procedure described in Ref. [5]. This showed the amplitude of fluctuations recorded at C

DOI: 10.1051/e3sconf/20161202003

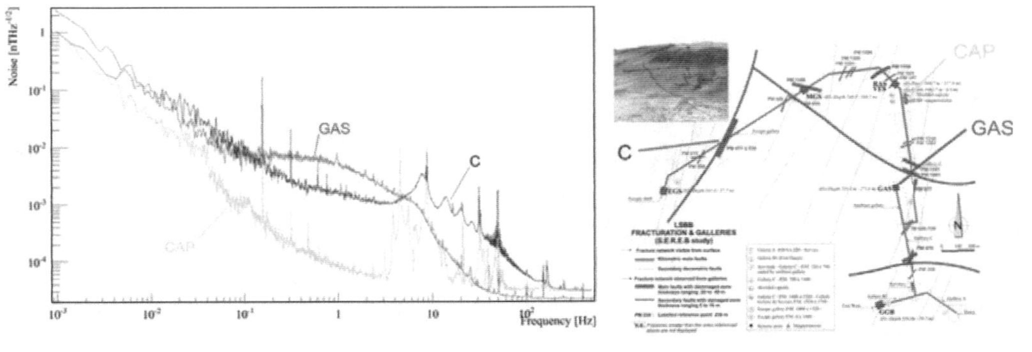

Figure 1. Left: frequency spectra of geomagnetic fluctuations recorded at different locations across the LSBB tunnel network: the shielded capsule (CAP), the GAS tunnel, and water flow point C. Right: a map of the LSBB laboratory showing the locations of the three measurement points.

Table 1. Scaling factors to convert the signal recorded by the $[\text{SQUID}]^2$ system to fit that recorded by the other SQUID systems.

Magnetometer and Location	Amplitude relative to $[\text{SQUID}]^2$
SQUID #1 C	1.57±0.16
SQUID #1 CAP	1.14±0.07
SQUID #2 GAS	1.61±0.14
SQUID #2 CAP	1.15±0.06

and GAS was 1.6 times that in the Capsule. When the SQUIDs used for these measurements were tested in the Capsule they recorded a 15% higher signal, which could be due to an error on the calibration, or the strong field gradients inside the metal capsule. We conclude there is no significant difference between the GAS and C, but the additional shielding at the Capsule reduced the amplitude of quasi-dc fluctuations by 40%. This is compatible with earlier measurements [4].

3. Tracking magnetic fields: SQUIDs versus fluxgates

The field at point C was monitored both with a SQUID system and a 3-axis fluxgate magnetometer. This provided an opportunity to compare the two instruments. The fluxgate sensor used was a Bartington Instrument *Mag03MSL100* low noise sensor (in April) and a *Mag661* sensor (in October). These sensors output three voltages proportional to the absolute value of the three magnetic field components. To monitor the field to the full (sub-nT) resolution, this was amplified by a factor of 100 (after subtracting the DC offset) before digitization using electronics build in-house and integrated with the SQUID readout.

Comparing the SQUID and fluxgate signals allowed us to check for drift of the fluxgate signals. It was seen that when initially turned on the signal sometimes showed some drift for the following hours, but after this, it was stable for long periods and followed the same pattern as the SQUID signal as shown in Fig. 2 (left).

Figure 2 (right) shows the frequency spectra of the signals recorded by the two sensors. The SQUIDs can achieve a reduction in the noise level typically ~10 (depending on frequency), revealing phenomena such as the Schumann resonances. The two lines in come together at very low frequencies. For long term tracking, the fluxgates are perfectly adequate as the magnitude of geomagnetic fluctuations is much greater than the instrument noise.

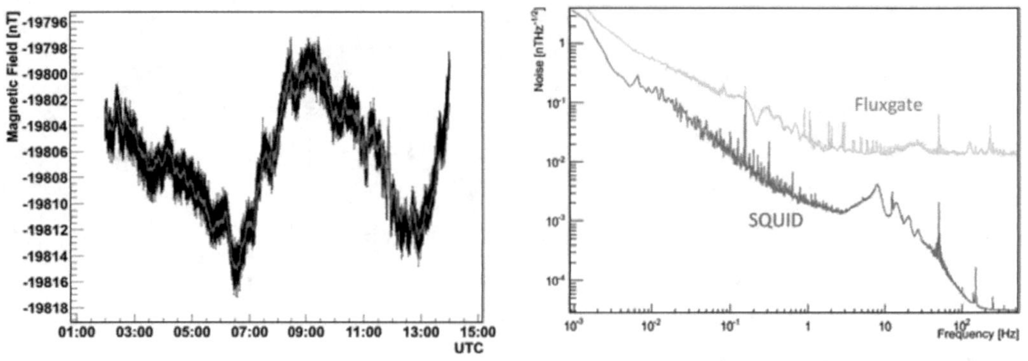

Figure 2. Left: magnetic signals recorded by the fluxgate (black line) and SQUID (red line) over a fourteen hour period. The 3-axis SQUID data has been transformed to fit this channel of the fluxgate to correct for difference in alignment of the two sensors. Right: frequency spectra of geomagnetic fluctuations recorded at point C using the SQUID (red) and fluxgate (green).

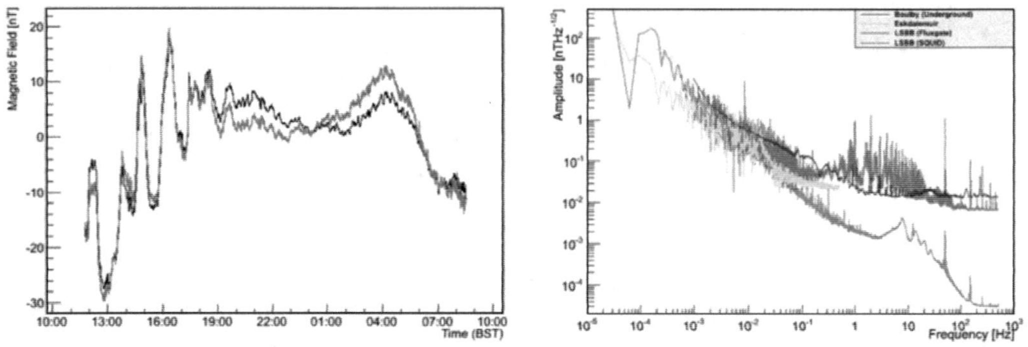

Figure 3. Left: magnetic signals recorded by fluxgate sensors at the Boulby Underground Laboratory (black line) and the INTERMAGNET data from the Eskdalemuir geomagnetic observatory, scaled to fit the Boulby data (red line). Right: frequency spectra of magnetic signals recorded at the Boulby Underground Laboratory (red), at LSBB using SQUIDs (pink) and fluxgates (blue). The data was not recorded synchronously. The spectrum of fluctuations recorded at Eskdalemuir is also shown (green).

Fluxgates have the advantage that they do not suffer from flux jumps and track the absolute field.

This study raises the possibility of combining the data sets from the two sensors. SQUID flux jumps could be more accurately identified and corrected by using the fluxgate data. Thus we could (in principle) produce a data set combining the resolution of the SQUID with the long term stability and absolute measurement provided by the fluxgate.

4. Comparisons with Boulby Underground Laboratory

The Boulby Underground Laboratory is a research facility 1.1 km underground in a potash mine in Yorkshire, England. Originally used for astroparticle physics experiments, it is now a general purpose underground laboratory.

We carried out a series of short term tests at Boulby using *Mag03MSL* fluxgate sensors to assess the suitability of the laboratory for geomagnetic measurements and the potential of magnetotelluric measurements in such a location. To our knowledge, this is the first publication of geomagnetic signals recorded in such a location.

Figure 3 shows that with this setup we could monitor the geomagnetic fluctuations to a similar resolution (although not as accurately) as the Eskdalemuir geomagnetic observatory.

We carried out simultaneous measurements on the surface with the aim of studying the attenuation of these at the underground location. Unfortunately the magnetic noise from human activities around the surface site (such as road traffic) was much greater than the geomagnetic fluctuations.

Figure 3 (right) shows the frequency spectrum of this data. The noise level recorded at Boulby is the same order of magnitude as that recorded using fluxgates at low noise locations. This suggests that Boulby is a potentially good site for geomagnetic monitoring. Further tests using a SQUID system would help assess this. Unlike LSBB, Boulby was not set up for low electromagnetic noise, therefore it may be this could be further improved.

5. Conclusions

This paper outlines the motivation for, and details of magnetic field measurements carried out using three separate SQUID magnetometers tracking geomagnetic fluctuations at different locations within the LSBB laboratory in a simultaneous operation.

We monitored the quasi-dc magnetic fluctuations at the C and GAS points and compared this to the data recorded by the $[SQUID]^2$ system in the capsule. Frequency spectra show that the Schumann resonances are clearly visible at point C.

These studies have demonstrated the potential of using multiple SQUID stations spread across LSBB to study magnetotelluric effects, search for local signals beneath the geomagnetic background, and investigate other phenomena.

This project has been possible thanks to the support of the Science and Technology Facilities Council (Follow on Fund grant ST/L001934/1), INRA-UAPV-EMMAH; the Oxford-CNRS collaboration scheme; and Hertford College, Oxford. The authors would like to thank all the staff at LSBB for their assistance, Prof Hans Kraus (Oxford) who developed the data acquisition hardware and software for the Oxford system, and Bartington Instruments for the loan of two fluxgate sensors. We also thank the staff at the Boulby Underground Laboratory for their assistance. The results shown include data collected at the Eskdalemuir geomagnetic observatory. We thank the British Geological Survey, for supporting its operation and INTERMAGNET for promoting high standards of magnetic observatory practice (`www.intermagnet.org`).

References

[1] S. Gaffet et al. Geophys. J. Int. **155**, 981–990 (2003).
[2] S. Henry et al. Nucl. Instrum. Meth. A **763**, 483–494 (2014).
[3] S. Henry et al. J. Instrum. **3**, P11003 (2008).
[4] V. Andrieux et al. i-DUST 2010, 02003 (2011).
[5] S. Henry, E. Pozzo di Borgo, and A. Cavaillou, Rev. Sci. Instrum. **84**, 024501 (2013).
[6] S. Henry et al. E3S Web of Conferences **4**, 02004 (2014).
[7] A.S. Orange, Proc. IEEE **77**, 287 (1989).
[8] A. Chave and A. Jones, The Magnetotelluric Method, CUP 2012.
[9] V. Korepanov and A. Marusenkov, Surv Geophys **33**, 1059–1079 (2012).
[10] T.D. Gamble, W.M. Goubau, and J. Clarke, Geophysics **44**, 53–68 (1979).
[11] C. Kwisanga. SQUID Geomagnetic Signal Analysis and Modelling of Schumann Resonances in the Earth-Ionosphere Cavity (Ph.D. thesis) Stellenbosch University (2016) `https://scholar.sun.ac.za/bitstream/handle/10019.1/98597/kwisanga_geomagnetic_2016.PDF`.

Characteristics of lightning flashes generating sprites above storms

Serge Soula[1], Oscar van der Velde[2], Joan Montanya[2], Martin Fullekrug[3], Andrew Mezentsev[3], and Janusz Mlynarczyk[4]

[1] Laboratoire d'Aérologie, Université de Toulouse, CNRS, Toulouse, France
[2] Electrical Engineering Department, Technological University of Catalonia, Terrassa, Spain
[3] University of Bath, Department of Electronic and Electrical Engineering, Bath, UK
[4] Department of Electronics, AGH University of Science and Technology, Krakow, Poland

Abstract. Sprites are Transient Luminous Events (TLEs) that can extend vertically from 40 to 90 km and horizontally over several tens of km to form clusters of individual or multiple column or/and carrot-shaped luminous elements. They can even extend over more than 100 km in the form of sequential luminous emissions that are called "dancing sprites". Their optical detection and other parameters describing the storm and the lightning activity associated allow us to understand the conditions of their production and their links with the lightning activity. Our observations confirm some characteristics of the sprites and put forward others: (i) the sprites are essentially produced above the stratiform region of the Mesoscale Convective Systems after positive cloud-to-ground lightning flashes that produce large Charge Moment Change (CMC), with a shorter delay if the impulsive CMC (iCMC) is larger. (ii) The dancing sprites reflect the timing and the location of the successive lightning strokes that generate them. (iii) The sprite elements can be shifted from the stroke location when their delay is large. (iv) Bright sprites produce current signatures in ELF radiation a few milliseconds (<5 ms) after the positive strokes that generate them.

1. Introduction

Sprites constitute a type of Transient Luminous Events (TLEs) often observed since the first dedicated campaigns in the US [1–3]. They have various shapes structure of streamers that classify them in several types, e.g., columnar, carrot, jellyfish, angel... [4–7]. Sprites can also horizontally extend over several tens of km in the form of sprite clusters [3, 8, 9] and they can even occur over more than ∼100 km in the form of sequential luminous emissions that are called "dancing sprites" [10, 11].

Most of the time, sprites occur above stratiform regions of Mesoscale Convective Systems (MCS), shortly after sprite-producing positive cloud-to-ground (SP+CG) strokes that lower a large amount of charge to the ground [2, 12–14]. Thus, after the SP+CG stroke, the sprite discharge develops as a streamer as a result of the strong transient electrostatic field that exceeds the threshold for dielectric breakdown in the middle atmosphere, generally around 70 km [15]. The time delay after the lightning stroke ranges from a few ms to several tens of ms [13, 16]; the time delay depends on the characteristics of the SP+CG stroke, especially in terms of CMC, iCMC and current waveforms [17]. It can reach a few tens of ms after a long lightning continuing current following the stroke [18].

Generally, the SP+CG flashes start close to the convective cores of the storm and horizontally extend into stratiform region that consists of several layers of alternating charge polarity at different altitudes. These flashes supposedly follow trajectories of charged ice particles toward the stratiform region of the storm system [19, 20]. Some of these flashes can be considered as spider lightning [21, 22]. In a recent work [20] the authors identified different modes of evolution of the negative leaders throughout the thundercloud system that can explain the positioning and timing of the positive +CG strokes. Thus, the negative leaders may propagate at various altitudes with different velocities and over various distances, while generating several multiple +CG strokes.

This paper summarizes the characteristics of the lightning flashes that produce sprites and the different links between both. Data from several case studies enable us to illustrate the findings collected during the last years in this contribution.

2. Data

The videos used in this study are recorded with low-light Watec 902H cameras oriented to the storm with a pan-tilt unit that can be remotely controlled via the Internet [17]. They operate in a triggered mode provided by the UFOCaptureV2 software from Sonotaco Company, to capture luminous events with brightness above a given threshold. The obtained video imagery has a time resolution of 25 frames (or 50 interlaced fields) per second, which corresponds to a time resolution of 20 milliseconds. The video camera also records GPS-referenced time and inserts the time in each video frame. The azimuth and elevation of the sprite events are determined with the software "Cartes du Ciel" (SkyCharts) by overlaying known stars on an image from the video imagery.

Lightning location systems monitor the lightning flash activity and provide most of SP+CG strokes within the studied area. The European lightning detection network operated by European Cooperation for Lightning Detection (EUCLID) records CG flash characteristics such as the location, polarity, peak current, and the occurrence times of CG strokes. The sensors of the network use both magnetic direction finding (MDF) and time of arrival (TOA) techniques to determine the location of CG strokes [18]. The detection efficiency (DE) of this network is ~90% over land and close to the coastline. The total lightning activity can be documented in restricted areas, as for example in Catalonia with the Lightning Mapping System XDDE operated by the Meteorological Service of Catalonia (SMC). During the field experiment of HyMeX at the Fall 2012, a Lightning Mapping Array was installed in southeastern France [17]. These systems detect and locate very high frequency (VHF) sources produced by different sections of a lightning leader, by interferometry for one and by difference of time of arrival for the other.

A system located at University of Bath (51.71N; 2.32W) records broadband ELF/VLF/LF/MF electromagnetic waves. It consists of a metal plate insulated from the ground to measure the vertical electric field, a precise GPS clock for timing the data acquisition, and an analogue signal conditioning and digital data acquisition unit [23]. This instrument has the capability to record electric field magnitudes in the frequency range from approximately ~4 Hz to ~400 kHz with a sampling frequency of 1 MHz, an amplitude resolution of $\sim 35\,\mu\mathrm{V\,m^{-1}}$, and a timing accuracy of ~10–20 ns.

The charge moment change (CMC), the impulsive CMC (iCMC) and the current moment waveform associated with the CG flashes [13] are reconstructed from ELF receiver measurements, by using the method presented by Mlynarczyk et al. [24]. The ELF recordings are performed by the Hylaty ELF station in Poland [25].

Reflectivity from meteorological radar is used to describe the structure of the storm. 24 Doppler radars (C band and S band) cover the entire area of mainland France and according to the location of the storm system analyzed, data from several radars can provide the 3D fields of radar reflectivity every 15 minutes up to a height of 14 km amsl. The

 DOI: 10.1051/e3sconf/20161202001

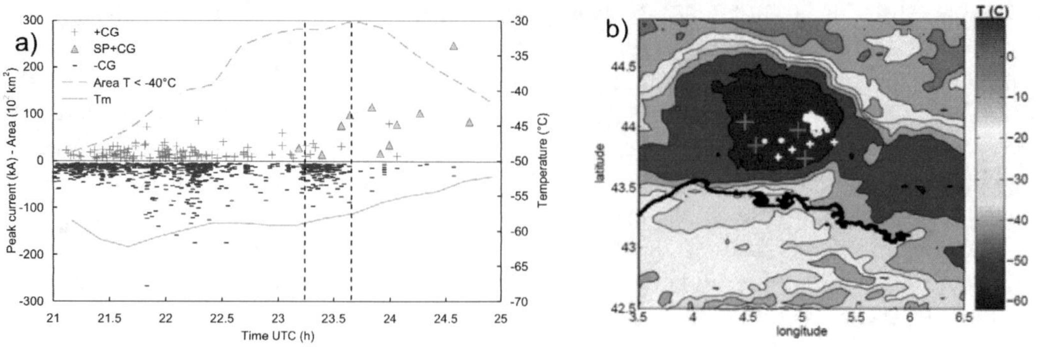

Figure 1. a) Time series of the peak current of +CG, SP+CG and −CG strokes, minimum temperature of the cloud top Tm, and of the cloud area with a temperature T < −40 °C for the storm on October 22nd of 2012. b) Distribution of the cloud top temperature in the area of the storm at 2325 UTC and location of the +CG (white cross), SP+CG (red plus) and –CG (white circle) strokes between 2315 and 2335 UTC.

horizontal (resp. vertical) resolution of reflectivity fields is 1 km (resp. 0.5 km). The cloud top temperatures are provided by the Meteosat satellite from European Organization for the Exploitation of Meteorological Satellites (EUMETSAT). Meteosat Second Generation (MSG) Spinning Enhanced Visible and Infrared Imager (SEVIRI) is based on radiometer data in the thermal infrared band (IR) at 11–13 μm. The parallax error is taken into account for the plots which associate the locations of CG strokes and their parent clouds.

3. Sprite and storm activity

The activity and the sprites produced by a storm in southeastern France on October 22nd, 2012, could be monitored during its whole lifetime. Two cameras allowed us to detect 12 sprite events and several lightning detection systems provided the total lightning activity, especially thanks to a LMA operated during the HyMeX campaign [17]. The storm is classified as a MCS with an activity of ~4 hours. Figure 1a displays the time series of the CG strokes characterized by their peak current, the size of the storm system (area T < −40°C) and the minimum value of the cloud top temperature. The SP+CG strokes occur during the last third of the storm activity and start when the CG flash rate decreases drastically, the size of the system is maximum and well after the minimum of the cloud top temperature (CTT), i.e. after the strong convection. All strokes preceding a sprite are positive (SP+CG) and their peak current is much larger in average than that of the other +CG strokes. Figure 1b displays the distribution of the CTT at 2325 UTC and the location of the lightning strokes detected between 2315 and 2335 UTC, including four SP+CG strokes. The latter are spatially separated from most −CG strokes that cluster in a specific area corresponding to the convective region [17]. Thus this sample confirms the typical location of the SP+CG strokes within the stratiform region of the MCS.

4. Sprite and lightning geometry

Figure 2 displays a case of lightning flash producing a sprite triangulated from two cameras detections (Fig. 2c) during the storm described in Sect. 3. In Fig. 2a and 2b, the VHF sources localized in 3D thanks to a LMA are plotted versus time with the longitude and the altitude, respectively. Figure 2d displays their 2D location within the storm system characterized by the radar reflectivity at 3000 m altitude. The flash starts close to the large reflectivity of the

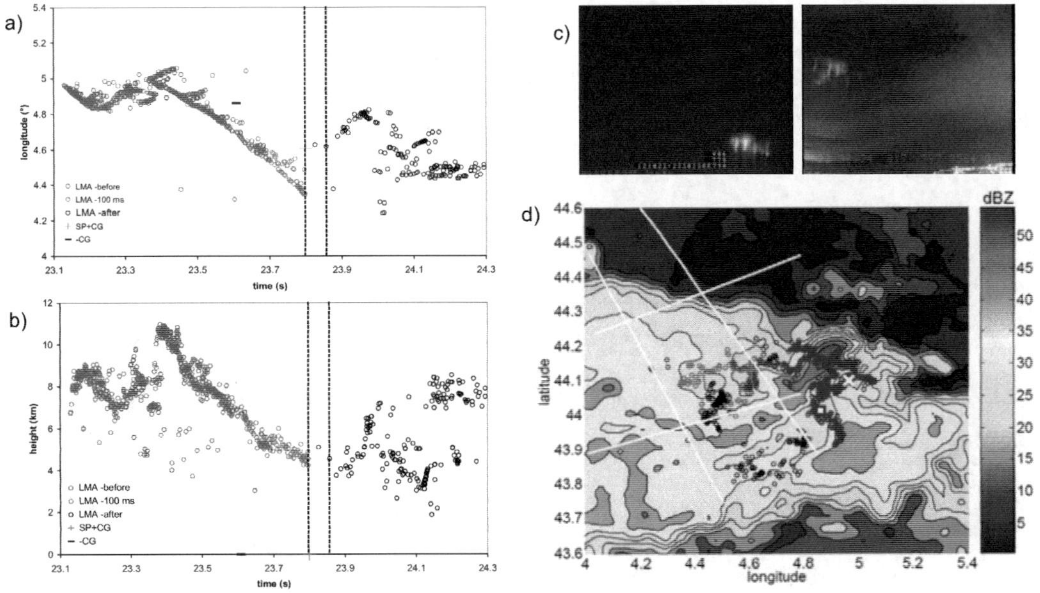

Figure 2. a) and b) Time series of the longitude and the altitude of the LMA-detected VHF sources (blue/red/black for at least 100 ms before/less than 100 ms before/after sprite event) and Euclid-detected CG strokes, after 2323 UTC. c) Two frames of the same sprite issued from two cameras at t = 23.798 s and t = 23.797 s, respectively. Distribution of the CTT in the area of the storm at 2325 UTC and location of the +CG (white cross), SP+CG (red plus) and −CG (white circle) strokes between 2315 and 2335 UTC. (d) Radar reflectivity field at the altitude of 3000 m at 2345 UTC, with superimposed the same events of the lightning activity as in a) and b). The white cross indicates the first VHF source detected and the white lines display the lines of sight of the main sprite elements.

convective region (white cross), remains in the same area for ~0.25 s and propagates westerly to develop its main branch (blue dots) at an altitude lower and lower across the stratiform region. It continues its propagation at a speed ~1.5 10^5 m s^{-1}, typical of a negative leader, until the sprite triggers (red dots). The SP+CG stroke (red plus) is located at the rear of the leader, within the stratiform region and within the same video frame (white lines in Fig. 2d) that is used for the triangulation of the sprite cluster.

5. Sprite and lightning chronology

A case of dancing sprite was detected during a storm on October 29th of 2013, over the Mediterranean Sea in the region of Balearic Islands. This sprite was especially long in time with four sequences of light emission, each one associated with a SP+CG stroke, as illustrated in Fig. 3b. Indeed, the sequences are indicated with horizontal lines that correspond to the four sequences illustrated with two frames for each in Fig. 3a. The peak current indicates the occurrence of the SP+CG strokes that coincide exactly with the beginning of the sequences, except for the first one with a long time-delayed sprite (~70 ms). Figure 3b displays other information of lightning activity, especially IC activity (cross at an arbitrary value of 100 kA) that occurs either during or before the sprite sequences. The electric field radiated in the VLF/LF range corresponds to the SP+CG strokes (Fig. 3c).

6. Sprite and lightning current waveform

The SP+CG strokes can be characterized by the peak current as it is seen in the previous sections, but also by the CMC, the iCMC and the current moment waveform. This information

 DOI: 10.1051/e3sconf/20161202001

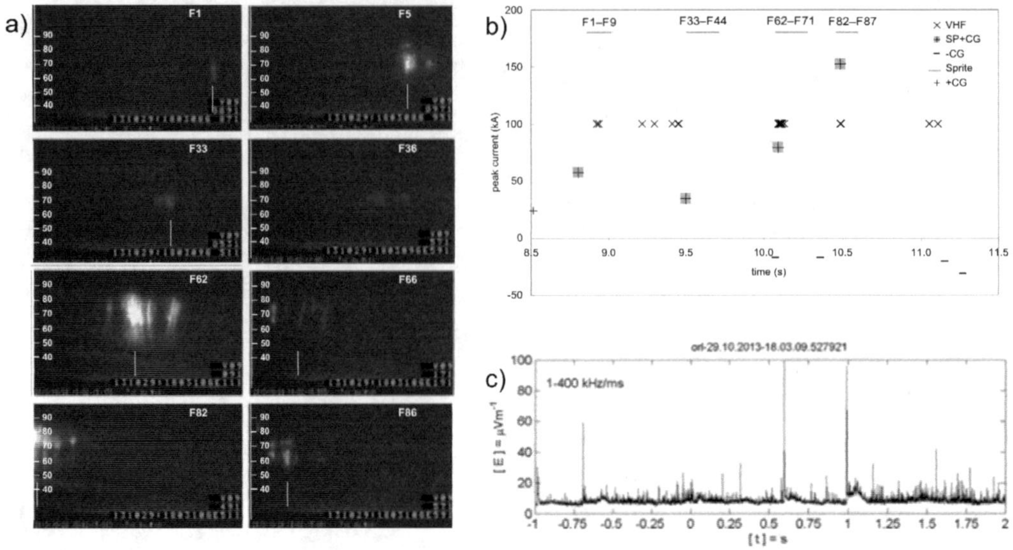

Figure 3. Case of dancing sprite at 1803 UTC on October 29th of 2013: a) 8 specific frames issued from the video imagery numbered from F1 that is the first frame with sprite luminosity. (b) Peak current versus time for 3 seconds for −CG strokes (black minus), +CG strokes (black plus), SP+CG strokes (large plus) and IC strokes plotted with an arbitrary current value of 100 kA (black cross). Four sequences of sprite luminosity are indicated with grey lines and frame numbers. (c) Electric field radiated in VLF/LF range for the same period of 3 seconds (t = 0 corresponds to t = 9.528 s in upper graph).

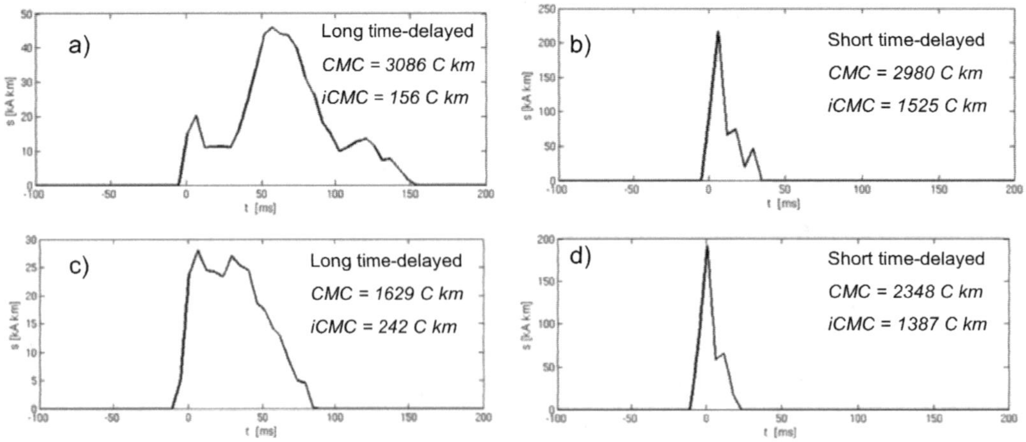

Figure 4. Current moment waveform for 4 SP+CG strokes recorded on October 22nd of 2012: a) and c) correspond to long-delayed time sprites; b) and d) correspond to short-delayed time sprites. The time t = 0 corresponds to the SP+CG stroke detection.

describes how the charge is lowered in the lightning channel. The graphs of Fig. 4 display four cases of SP+CG strokes that illustrate the observations made in different cases of storms. a) and c) correspond to long time-delayed sprites while b) and d) correspond to short time-delayed sprites. Furthermore, b) and d) correspond to bright sprites. Thus, it is observed that the +CG strokes with a low iCMC can lead to a sprite emission but with a long delay that allows more charge to be neutralized during the lightning process. The short time-delayed sprites are triggered by SP+CG strokes with a strong iCMC, i.e. a large amount of charge is rapidly neutralized by the lightning process. In all cases of sprites, the CMC of the lightning stroke has a value above a threshold of a few hundreds of C km. The brightness of the sprites also seems to depend on the iCMC of the parent stroke.

7. Conclusion

Several case studies of sprite-producing storms allow us to characterize the SP+CG strokes, and in a more comprehensive way, the flashes that include them. The sprite-producing lightning strokes are positive and located in the stratiform region of the storm. They occur at the end of the storm activity when the stratiform region is well developed. The SP+CG flashes start generally close to the convective region and follow long paths across the stratiform region. The SP+CG strokes produce large CMCs, while their iCMC can be either low and the sprites occur after a long delay, or large and the sprites occur rapidly and very bright. Dancing sprites are produced by successive SP+CG strokes or pulse currents associated with the lightning process (continuing current, M component…). The bright sprite can provide a substantial contribution to the current signature.

References

[1] D.D. Sentman, E.M. Wescott, Geophys. Res. Lett. **20**, 24 (1993).
[2] W.A. Lyons, J. Geophys. Res., **101**, D23 (1996).
[3] D.D. Sentman, E.M. Wescott, D.L. Osborne, D.L. Hampton, M.J. Heavner, Geophys. Res. Lett., **22**, 10 (1995).
[4] W.A. Lyons, T.E. Nelson, E.R. Williams, S.A. Cummer, M.A. Stanley, Mon. Weather Rev., **131**, 2417 (2003).
[5] H.C. Stenbaek-Nielsen, M.G. McHarg, J. Phys. D Appl. Phys., **41**, 234009 (2008).
[6] J. Li, S.A. Cummer, Geophys. Res. Lett., **36**, L10812 (2009).
[7] J. Montanyà, O. van der Velde, D. Romero, V. March, G. Solà, N. Pineda, M. Arrayas, J.L. Trueba, V. Reglero, S. Soula, J. Geophys. Res., **115**, A00E18 (2010).
[8] M. Füllekrug, D.R. Moudry, G. Dawes, and D.D. Sentman, J. Geophys. Res., **106,** D17 (2001).
[9] S. Soula, F. Iacovella, O. van der Velde, J. Montanyà, M. Füllekrug, T. Farges, J. Bór, J.-F. Georgis, S. NaitAmor, J.-M. Martin, Atmos. Res., **135**, 415–431 (2014).
[10] J.R. Winckler, W.A. Lyons, T.E. Nelson, R.J. Nemzek, J. Geophys. Res., **101,** D3 (1996).
[11] M. Füllekrug, A. Mezentsev, S. Soula, O. van der Velde, T. Farges, Geophys. Res. Lett., **40**, 2395–2399 (2013).
[12] D.J. Boccippio, E.R. Williams, S.J. Heckman, W.A. Lyons, I. Baker, R. Boldi, Science, **269**, 1088–1091 (1995).
[13] S.A. Cummer, W.A. Lyons, J. Geophys. Res., **110**, A04304 (2005).
[14] S. Soula, O. van der Velde, J. Montanyà, T. Neubert, O. Chanrion, M. Ganot, Atmos. Res., **91**, 2–4 (2009).
[15] V.P. Pasko, U.S. Inan, T.F. Bell, Y.N. Taranenko, J. Geophys. Res., **102**, A3, 4529 (1997).
[16] O.A. van der Velde, Á. Mika, S. Soula, C. Haldoupis, T. Neubert, U.S. Inan, J. Geophys. Res., **111**, D15203 (2006).
[17] S. Soula, et al., J. Geophys. Res. Atmos., **120**, 11,552 (2015).
[18] S.A. Cummer, M. Füllekrug, Geophys. Res. Lett., **28**, 3 (2001).
[19] L.D. Carey, M.J. Murphy, T.L. McCormick, N.W.S. Demetriades, J. Geophys. Res., **110**, D03105 (2005).

[20] O.A. van der Velde, J. Montanya, S. Soula, N. Pineda, J. Mlynarczyk, J. Geophys. Res. Atmos., **119** (2014).
[21] V. Mazur, X. Shao, P.R. Krehbiel, J. Geophys. Res., **103**, D16 (1998).
[22] T.J. Lang, S.A. Rutledge, K.C. Wiens, Geophys. Res. Lett., **31**, L10105 (2004).
[23] M. Füllekrug, Meas. Sci. Technol., **21** (2010).
[24] J. Mlynarczyk, J. Bór, A. Kulak, M. Popek, J. Kubisz, J. Geophys. Res. Space Physics, **120**, 2241 (2015).
[25] A. Kulak, J. Kubisz, S. Klucjasz, A. Michalec, J. Mlynarczyk, Z. Nieckarz, M. Ostrowski, S. Zieba, Radio Sci., **49**, 361 (2014).

First analyses of the iOSG-type superconducting gravimeter at the low noise underground laboratory (LSBB URL) of Rustrel, France

Séverine Rosat[1,a], Jacques Hinderer[1], Jean-Paul Boy[1], Frédéric Littel[1], Daniel Boyer[2], Jean-Daniel Bernard[1], Yves Rogister[1], Anthony Mémin[3], and Stéphane Gaffet[2]

[1] Institut de Physique du Globe de Strasbourg, UMR 7516, Université de Strasbourg/EOST, CNRS, 5 rue Descartes, 67084 Strasbourg, France
[2] LSBB URL, France
[3] Université Côte d'Azur, CNRS, Observatoire de la Côte d'Azur, IRD, Géoazur, France

Abstract. In the last few years, the performance of the cryogenic gravity instruments has been further improved by the development of a new generation of superconducting gravimeter (SG): the so-called iOSG which is a superconducting gravimeter designed for observatory purpose with a heavier sphere than previous SGs. The first iOSG (iOSG-024) has been installed in July 2015 at the LSSB low background noise underground research laboratory in Rustrel (France), funded by the EQUIPEX MIGA (Matter wave-laser based Interferometer Gravitation Antenna) project and by the European FEDER 2006-2013 "PFM LSBB – Développement des qualités environnementales du LSBB". This instrument is operational since September 2015. We present the first tidal analyses of the 7-month time-varying gravity records of this newly installed instrument as well as the calibration results performed by parallel FG5 absolute gravity measurements. We also show the performances of iOSG-024 in terms of noise levels in the seismic (in the millihertz frequency range) band using a standardized procedure based on the computation of the residual power spectral densities over a quiet time period. The obtained noise levels are compared with other SG sites and with seismological reference noise models. The combination of the instrumental performance of the iOSG with the LSBB site properties makes this gravimetric station one of the quietest in the world, comparable to the lower sensor of the OSG-56 at BFO, at seismic frequencies.

1. Introduction

Superconducting Gravimeters (SGs) are relative instruments that continuously measure the time-varying gravity field at the Earth's surface. Since the development of the first commercialized SGs [1] in the 90s, the size and instrumental noise have been reduced [2]. In the last few years, the performance of the cryogenic gravity instruments has been further improved by the development of a new generation of superconducting gravimeter: the so-called iOSG which is a superconducting gravimeter designed for observatory purpose. A special SG prototype (called OSG-056), with sensor properties intermediate between the compact SG and the new iOSG, has already been recording since 2009 at the Black Forest

[a] Corresponding author: severine.rosat@unistra.fr

DOI: 10.1051/e3sconf/20161206003

Figure 1. The iOSG-24 recording at the LSBB in Rustrel (France) since July 2015.

Observatory (BFO, Germany) with promising results in terms of sensitivity [2]. The OSG-56 recording at BFO is indeed a dual-sphere instrument with a higher sphere weighting 4.3g and a lower sphere which is heavier with a mass of 17.7g. So the lower sensor of OSG-56 (noted OSG-56_L) is similar to the iOSG. The first iOSG (iOSG-024; Fig. 1) has been installed at the LSSB low background noise underground research laboratory (URL) in Rustrel (France), funded by the EQUIPEX MIGA (Matter wave-laser based Interferometer Gravitation Antenna, [3]) project and was installed in July 2015. First data have been recorded since mid-September 2015.

The SG measurement relies on the magnetic feedback force of a levitated Niobium sphere in the vacuum. In order to keep the superconductivity of the sphere, the sensor is operating in a liquid Helium bath maintained at 4.2 K. A liquefying system enables to recycle the Helium gas into liquid Helium making the human intervention for liquid Helium refilling unnecessary. A known source of anthropogenic noise is hence reduced.

The purpose of increasing the mass of the levitated sphere is to decrease the thermal noise due to Brownian motion inside the sensor. The spectral acceleration-noise power density ($P_{thermal}$) due to Brownian motion in a simple mechanical oscillator is given by:

$$P_{thermal} = 4k_B T \frac{\omega_0}{m Q} \tag{1}$$

where ω_0 is the natural frequency of the oscillator, Q its quality factor and m is the mass of the oscillating sphere [4]; k_B is the Boltzmann constant and T the temperature.

First, in Sect. 2, we show the calibration results. Next, in Sect. 3, we plot the noise levels at the seismic frequencies to compare the performance of the iOSG-24 with other worldwide SG stations. Finally, in Sect. 4, we present the noise levels in the tidal frequency bands.

2. Calibration

A usual way to calibrate relative gravimeters is to perform simultaneous absolute gravity measurements during a few days and to estimate a scale factor by least-squares adjustment of the tidal amplitudes (see e.g. [5]). This parallel FG5 measurement was realized on 5 days from October 8 to 12, 2015 with the FG5#206 (serial number 206) absolute gravimeter owned by the Strasbourg gravimetric observatory. The resulting superposition of absolute gravity measurements and relative SG time-varying records is represented in Fig. 2. The obtained scale factor value is - -451.2 nm/s^2 /V $\pm$ 0.27 nm/s^2 /V (relative error of 0.06%).

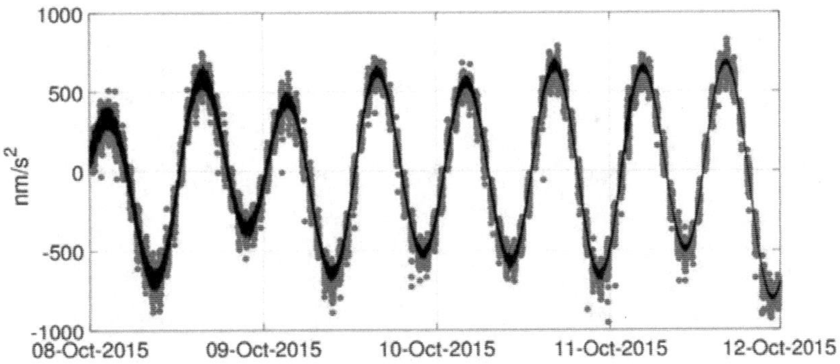

Figure 2. Superposition of the FG5#206 absolute drop values (in red) with the iOSG-24 (in black) recording at the LSBB in Rustrel (France) after applying a scale factor of -451nm/s^2 /V.

3. Seismic noise levels

A standard procedure was introduced by [6, 7] to compare noise levels of various gravimetric sites. Since other worldwide SG data are available at a sampling interval of one minute, we down-sample the raw one-second iOSG-24 data to one minute using a low-pass filter with a cut-off period of two minutes. Then we consider daily time windows and remove solid tides for an elastic reference Earth model from the data as in [7]. A local atmospheric pressure reduction is also applied using a nominal admittance of -3 nm/s^2 /hPa. The five quietest days in terms of RMS deviation are selected. Subsequently a Power Spectral Density (PSD) is computed using a smoothed periodogram estimate on these five quietest days.

The resulting PSD for the iOSG-24 is compared (Fig. 3) to some other noise levels at SG sites that were published in [2]. At frequencies larger than 1 mHz, the iOSG-24 has a noise level similar to the lower sensor of the BFO OSG-56 gravimeter which is among the lowest ones in the world. The OSG-56 is indeed a dual-sphere instrument where we denote by L the lower heavier sphere and by U the upper standard sphere. Below 1 mHz, the iOSG-24 has a noise level that becomes larger than BFO OSG-56_L. This could be due to the environment but further investigation is needed. Indeed, we cannot distinguish between environmental and instrumental noise with only one instrument at one site. When comparing the noise level of the lower sensor to the higher sensor of the BFO OSG-56, we see the efficiency of instrumental noise reduction by increasing the mass of the sphere (Eq. (1)) in the seismic frequency range considered here. However, it was shown in [2] that the noise reduction by increasing the mass of the sphere was not efficient any more below 0.1 mHz.

The high frequency decrease on Fig. 3 is due to the low-pass filtering that was applied before decimating data from one second to one minute.

A criterion of the quality of a site in terms of noise is also the observation of the incessant background free oscillations (the so-called "hum") [9], which are clearly visible on the iOSG-24 data at LSBB (Fig. 4). We have plotted the quartiles of the Global Seismographic Network noise levels [10] that represent a more recent reference noise model than the New Low Noise Model NLNM [8]. At frequencies larger than 1mHz, long-period seismometers have lower noise levels than the iOSG-24 (Fig. 4). Physical mechanisms to explain the incessant excitations of the fundamental seismic modes have been given by [11]. In fact, SGs are known to have a larger noise amplitude in the seismic band. This larger noise amplitude was partly indebted to the Brownian motion of the levitated sphere [12]. However, at frequencies smaller than 1 mHz, the SG has better performances due to a more efficient atmospheric pressure reduction [9].

DOI: 10.1051/e3sconf/20161206003

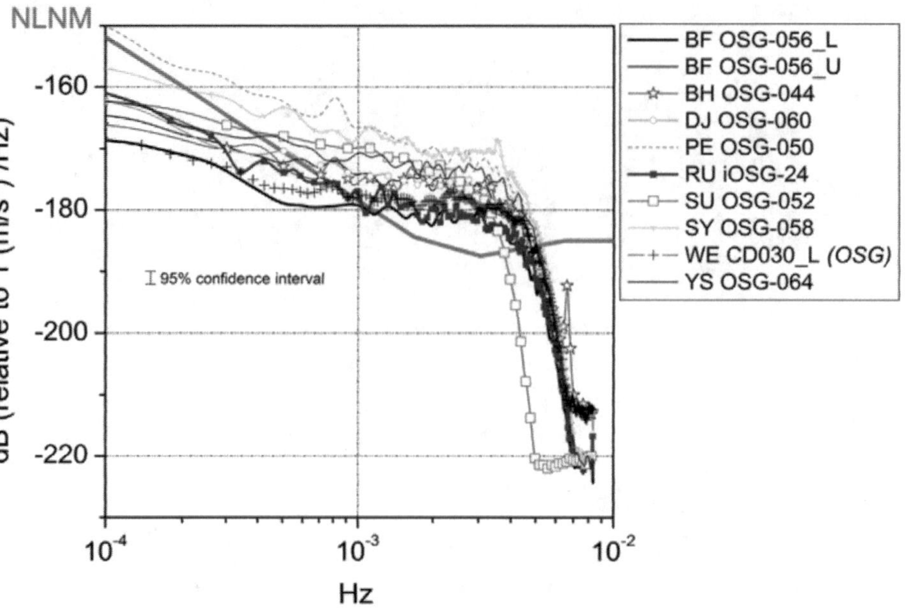

Figure 3. Noise level in the seismic band of the LSBB iOSG-24 site (named RU) compared with some worldwide SG sites: the dual-sphere OSG-56 at Black Forest Observatory (BF, lower -L- and upper -U- sensor, Germany), the OSG-60 at Djougou (DJ, Benin), the OSG-44 at Bad-Homburg (BH, Germany), the OSG-50 at Pecny (PE, Czech Republic), the OSG-52 at Sutherland (SU, South-Africa), the OSG-58 at Syowa (SY, Antarctica), the lower-sphere of the dual-sensor CD030 at Wettzell (WE, Germany) and the OSG-64 at Yebes (YS, Spain). The seismological noise model (NLNM) [8] is represented by the thick red line. The color version of this figure is available only in the electronic edition.

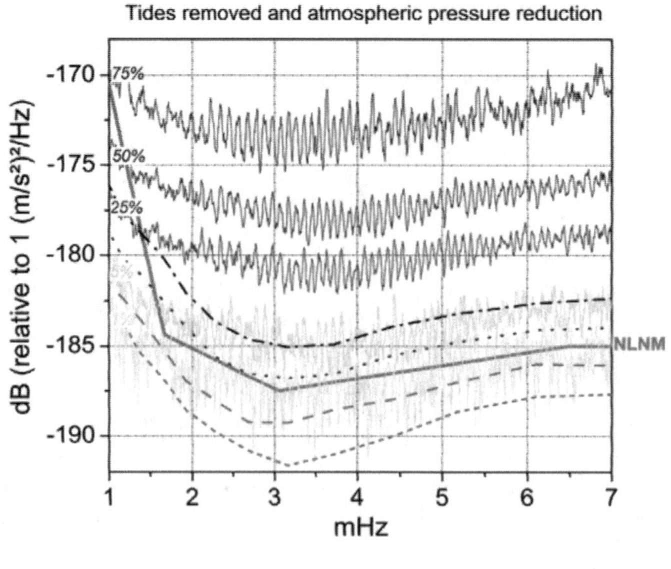

Figure 4. 1%, 5%, 25%, 50% and 75%-tiles of daily Power Spectral Densities of the LSBB iOSG-24 site (in cyan and blue lines) compared with the 1st, 5th, 25th and 50th percentiles of the Global Seismographic Network noise levels [10] respectively plotted in short-dashed gray, dashed gray, dotted gray and dot-dashed black lines. The seismological noise model (NLNM) [8] is represented by the thick red line. The color version of this figure is available only in the electronic edition.

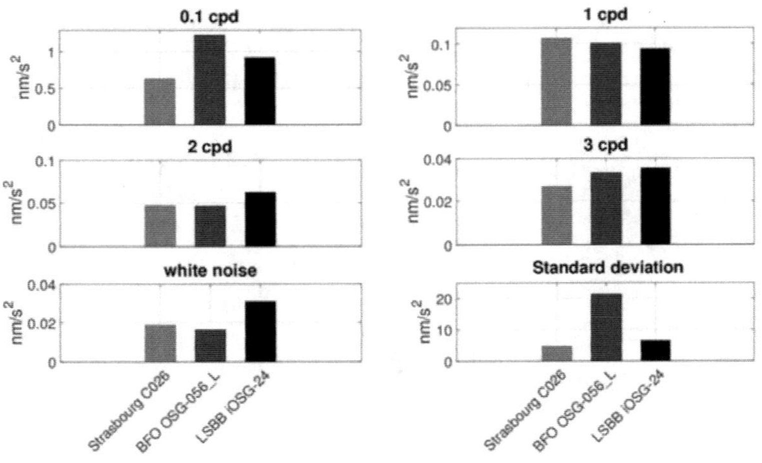

Figure 5. Average noise levels at frequency bands: 0.1, 1, 2, and 3 cpd. The corresponding white noise amplitude as well as the standard deviations are also given. These average noise levels are obtained after an ETERNA tidal analysis on 212 days of data. The color version of this figure is available only in the electronic edition.

4. Tidal analysis results

The analysis of tidal waves in a gravimetric time series can be performed with the ETERNA software [13] which enables to retrieve a tidal model at the nanogal precision (1 nGal = 0.01 nm/s^2). An ETERNA tidal analysis performs a least-squares adjustment based on a tidal potential catalogue (we choose the Hartmann & Wenzel catalogue [14]) to retrieve observed tidal amplitudes and phases, but also some noise levels in the tidal bands. These noise levels are the average FFT of gravity residuals from ETERNA analysis. They are normalized in order to be independent of the record lengths. The white noise estimate σ is obtained from the mean power spectral density (PSD) of the gravity residuals after ETERNA processing through the relation PSD $= \sigma^2\,T_0$, where T_0 is the sampling interval. The standard deviation of the gravity residual time series is also given.

The estimates of average noise levels in different frequency bands, 1, 2 and 3 cycle per day (cpd), are compared in Fig. 5 with the average noise levels obtained for the SG C026 recording at the gravimetric observatory of Strasbourg (France) and for the lower-sphere of the OSG-56 at BFO. We can see that the average noise level is slightly larger for the iOSG-24 at LSBB in the 2 and 3 cpd frequency bands than for the Strasbourg and BFO sites. Additional semi-diurnal and ter-diurnal tidal terms could be inserted in the least-squares adjustment in the tidal analysis in order to further improve the local tidal models and the removal of periodic terms at such frequencies. However, a longer time-series would be required, as we had only 212 days of records since the installation.

5. Conclusion

The environmental quiet site of the LSBB associated with the latest generation of superconducting gravimeters makes this gravimetric station one of the quietest in the world, particularly at seismic frequencies, with a noise level of −180 dB at 1 mHz. The high-quality data it can provide, combined with global data, are very promising in the search for small-amplitude geophysical signals.

The iOSG-24 gravimeter of the LSBB was funded by the EQUIPEX MIGA (Matter wave-laser based Interferometer Gravitation Antenna) ANR-11-EQPX-0028 and by the European FEDER 2006-2013 "PFM LSBB - Développement des qualités environnementales du LSBB".

References

[1] G. Goodkind, Phys. Rev., **84** (1999).
[2] S. Rosat, J. Hinderer, Bull. Seism. Soc. Am., **101**, No. 3, 1233–1241 (2011).
[3] B. Canuel, L. Amand, A. Bertoldi, W. Chaibi, R. Geiger, J. Gillot, A. Landragin, M. Merzougui, I. Riou, S.P. Schmid and P. Bouyer, E3S Web of Conferences, **4** (2014).
[4] R.J. Warburton, H. Pillai, and R.C. Reineman, Int. Association of Geodesy (IAG) Symposium Proceedings (2010).
[5] J. Hinderer, D. Crossley, R.J. Warburton, Treatise on Geophysics, **3** (2007).
[6] D. Banka, *Ph.D. Thesis*, GDMB-Informationgesellschaft mbH, Clausthal, Germany (1997).
[7] D. Banka, D. Crossley, Geophys. J. Int., **139**, 87–97 (1999).
[8] J. Peterson, U.S. Geol. Surv. Open-File Rept. 93-332, Albuquerque, New Mexico (1993).
[9] R. Widmer-Schnidrig, Bull. Seism. Soc. Am., **93**, No. 3, pp. 1370–1380 (2003).
[10] J. Berger, P. Davis, G. Ekström, J. Geophys. Res., **109**, B11307 (2004).
[11] K. Nishida, Geophys. J. Int., **199**, 416–429 (2014).
[12] Richter B., Wenzel H.-G., Zürn W. and Klopping F., Phys. Earth Planet. Inter., **91**, 131–148 (1995).
[13] H.-G. Wenzel, Bull. Inf. Marees Terr., **124**, 9425–9439 (1996).
[14] T. Hartmann, H.-G. Wenzel, Geophys. Res. Lett., **22** (24), 3553–3556 (1995).

E3S Web of Conferences **12**, 05001 (2016) DOI: 10.1051/e3sconf/20161205001
i-DUST 2016

Life in deep subsurface

Grégoire Galès[1,2,a] and Gaël Erauso[2]

[1] Aix-Marseille Université, CEREGE, Centre St Charles, Case 67, 3 place Victor Hugo, 13331 Marseille, France
[2] Aix-Marseille Université, Université du Sud Toulon-Var, CNRS/INSU, IRD, MIO, UM 110, 13288 Marseille Cedex 09, France

Abstract. Life extends far deeper into the Earth's subsurface than presumed possible 30 years ago. In the past, it was assumed that life is a surface phenomenon, and that even "hardy prokaryotic types" are not capable of living deeper than tens of meters below the surface [1]. In the 1990s, it became apparent that genetically and metabolically diverse microbial communities existed under highly reducing conditions in the deep subsurface [2]. Today we know that life in the deep subsurface is ubiquitous and comprises a large proportion of the biomass on Earth [3]. Many questions concerning life in the deep remain unanswered. What is the lower depth limit of the deep biosphere? Which energy sources are fueling these communities? How are genetic diversity and functional activity linked to geochemical factors? What we know is that the deep subsurface is an extreme environment and that the microorganisms living here have developed numerous mechanisms to deal with high pressure and temperature, limited energy and nutrient availability, extreme acidity and alkalinity, metal toxicity, and radioactivity [4].

1. Introduction

Prokaryotes are remarkable in their ability to harvest energy from myriad redox reactions [5]. Their metabolic diversity and plasticity allow them to colonize almost all natural and industrial environments. Redox and biochemical reactions catalyzed by bacterial enzymes can lead to mineral precipitation, dissolution or transformation. Therefore, bacteria affect the geochemistry of modern environments, and may have contributed to shaping the near-surface environment of the early earth. Subsurface is classically defined as terrestrial habitats below 8 m and marine sediments below 10 cm. Deeply buried subsurface rocks and sediments may harbour over half of all prokaryotic cells on earth [6]. In this short survey, we will estimate physicochemical constraints affecting microbial life in subsurface, describe tools used to study microorganisms, and then illustrate the previous section by presenting some microorganisms isolated from deep subsurface and some results obtained by molecular studies. We finally discuss about subsurface microbial heterogeneities.

[a] Corresponding author: gregoire.gales@univ-amu.fr

2. Physicochemical constraints on microbial life in subsurface

2.1 Nutrients

Bacterial energy supply comes from chemical sources, due to fluids that migrate in deep levels in the Earth. This supply lies in the oxidation of an electron donor (H_2, H_2S, CH_4, organic matter...) coupled with the reduction of an electron acceptor (CO_2, SO_4^{2-}, NO_3^-, O_2 when available...) as phototrophy is impossible in subsurface. Microorganisms can be distinguished into physiological groups according to the electron acceptor they use: SO_4^{2-} (sulfate-reducers), NO_3^- (nitrate-reducers), NO_2^- (nitrite-reducers), CO_2 (methanogens), FeIII (Fe-reducers) or O_2 (aerobic to microaerobic). Depending on their carbon source, microorganisms are said heterotrophs when their carbon supply comes from organic compounds, and autotrophs when their carbon supply comes from inorganic carbon. The type of metabolic processes that occurs in subsurface depends to a great extent on the availability of electron acceptors. Considering that stratal waters often contain sulfate at various concentrations, and carbonate, one can assume that the major metabolic processes in oil reservoirs are sulfate reduction, methanogenesis, fermentation, and possibly homoacetogenesis.

Sulfate reduction	$SO_4^{2}+ 4\,H_2+ 2\,H^+$	$->H_2S + 4\,H_2O$
Methanogenesis	$CO_2+ 4\,H_2$	$->CH_4+2\,H_2O$
Fermentation	$C_6H_{12}O_6$	$->2\,CO_2+ 2\,CH_3CH_2OH$
Homoacetogenesis	$2\,HCO_3^- + H^+ + 4\,H_2$	$->CH_3COOH + 4\,H_2O$.

The four corresponding bacterial groups are actually those having been the more frequently isolated from waters sampled from deep subsurface, and are generally considered as potentially indigenous to the subsurface formations. Hydrogen may be supplied for methanogenesis and sulfate-reduction by mineral hydrolysis, maturation of organic matter, metabolic activity of fermenters, or even from the oil aromatization [7]. Water radiolysis was a potential energy source for the deep subsurface biosphere according to [8]. H_2 constitutes a major component of dissolved organic gases (up to 98%) in the groundwater of Precambrian shields, and its concentration ranges up to several mM. These concentrations are several orders of magnitude higher than those observed in shallow aquifers [9]. Radiolysis consists in the decomposition of water due to the products of radioactive decay. Radiolysis has been proposed as a mechanism for generating large quantities of molecular hydrogen, this hypothesis being supported by the fact that H_2-bearing fluid inclusions in quartz are associated with U-bearing [10]. The H_2 yield under conditions that mimic natural radiolysis in subsurface consistent with millimolar H_2 concentrations found in Precambrian shields [8]. As no O_2 or H_2O_2, which are other products of water radiolysis were found in subsurface, they may have abiotically oxidized S^{2-} and Fe^{2+} derived from anaerobic microbial metabolism. This cycling mechanism would maintain the stability of anaerobic conditions as well as supply the electrons acceptors for anaerobic metabolism. The elevated H_2 abundances from radiolysis also enhance abiotic formation of hydrocarbons [11], lipids [12] and organic acids [13], which are thought to be produced by the reactions between CO or CO_2 with H_2 in the presence of metal sulfide catalysts.

Salinity, temperature, pressure and porosity are the main physical constraints on microbial life in subsurface. Bacteria and Archaea can for example inhabit unusual environments with low water availability. Inagaki [14] reported a biogeochemical and microbiological characterization of a microbial community inhabiting deep-sea sediments overlying a natural CO_2 lake, southern Okinawa Trough. They found high abundances ($>10^9\,cm^{-3}$) of microbial cells (mainly anaerobic methanotrophic archaea and sulfur-metabolizing bacteria) in sediment pavements above the CO_2 lake, decreasing to strikingly low cell numbers (10^7 cm^{-3}) at the liquid CO_2/CO_2-hydrate interface, despite of the potentially aggressive characteristics of this non-polar solvent.

2.2 Salinity

High salt concentrations increase water viscosity and density and decrease gases solubility by an effect called "salting-out". Ions are strongly hydrated, so they tie up a lot of water molecules. That means there is less water free to hydrate the dissolved gas, so less gas can remain dissolved in the salt solution. Microbial life can be found over the whole range of salt concentrations, from freshwater to hypersaline environments (salt lakes, salted food products, chotts, saltern ponds…), where they may reach high population densities [15]. Since biological membranes are permeable to water, cells cannot maintain the water activity of their cytoplasm higher than that of the surrounding environment, which would lead to a rapid loss of water to the environment. To cope with the osmotic stress, microbes developed two different adaptative strategies: the "salt in" strategy, in which cells maintain high intracellular salt concentration (at least equivalent to the external concentration), all intracellular systems being adapted to the presence of salts; the "compatible solute" strategy, in which cells maintain low salt concentrations in their cytoplasm by the action of ionic pumps. The osmotic pressure of the medium is then counterbalanced by organic compatible solutes (glycerol, glycine betaine, trehalose…) [15]. Therefore, life at high salt concentrations is energetically expensive since it involves the buildup and maintenance of steep ion concentration gradients across the cell membrane, whether or not accompanied by the biosynthesis or accumulation of organic osmotic compounds. A survey of the halophilic microorganisms shows that not all known metabolic types function in the presence of high salt concentrations. Denitrification, oxygenic and anoxigenic photosyntheses occur close to NaCl saturation, whereas autotrophic oxidation of NH_4^+ to NO_2^- does not seem to occur above 150 g of salt per liter. Methanogenesis is also sensitive to salt concentration, as reduction of CO_2 with H_2 and acetoclastic split are common in freshwaters and seem not to occur at high salt concentration. Homoacetogenesis and methanogenesis from methylamines seem to be prevalent in hypersaline environments.

2.3 Temperature

Temperature influence on microbial metabolism has been widely studied due to the biotechnological impact of enzymes coming from thermophile microbes. For example, the DNA polymerases of *Thermus aquaticus* or *Pyrococcus furiosus*, two hyperthermophiles, are widely used in PCR (Polymerase Chain Reaction) applications because of their stability when heating the PCR mixture. Microorganisms which inhabit high temperature environments are defined as thermophilic if their optimum growth temperatures are <45°C. If an organism has an optimum and maximum growth temperature of at least 80 and 90°C respectively, it is further defined as a hyperthermophile. Life at high temperature seems to appear in a wide diversity of environment. *Thermocrinis ruber*, an aerobic, facultatively chemolithotrophic bacterium that growths in the laboratory between 44 and 89°C by oxidizing hydrogen, elemental sulfur, thiosulfate, formate of formamide is hosted in an alkaline hot spring in the Yellowstone National Park, USA. Deep-sea hydrothermal systems at a depth of 2600 m on the East Pacific Rise support anaerobic autotrophic methanogens such as *Methanococcus jannaschi*, which grows optimally in the laboratory at 85°C. Acid solutions generated by interactions between volcanic gases and seawater in the Aeolian islands and the solfatara of Napoli (Italy) are the habitats of acidophilic Archaea, including *Acidianus infernus*, *Thermoplasma volcanium* and *Metallosphaera sedula*, which grow optimally at pH near 2.

2.4 Pressure

Pressure is a thermodynamic variable that varies from 1 to 1100 atmospheres (0,1 to 102 MPa) from the surface to the deep-sea bottom, and is up to 100 MPa in oil fields. Many thermophiles and hyperthermophiles are also barotolerant/barophiles (or piezophiles in recent terminology) and may employ metabolic processes that are affected by pressure. But at most conditions of biological interest, the effect of pressure on metabolic energetics is secondary to that of temperature [5]. Nevertheless, there is some evidence that elevated pressure may manifest interesting effects on cellular physiology [16]. Hydrostatic pressure in the range of 0.3–0.5 GPa usually inhibits the growth of microorganisms. DNA synthesis is one of the most pressure-sensitive cellular process, especially the initiation of DNA replication [17]. Protein synthesis is also highly susceptible to elevated pressure. Ribosomes associated with both mRNA and tRNA show perfect stability at pressures up to 102 Mpa, although uncharged ribosomes become dissociated at pressures >60 MPa [18]. Therefore, piezophiles are likely to have undergone critical changes in their ribosomal proteins. There is considerable evidence that an increased proportion of unsatured fatty acids in membrane lipids is associated with bacterial growth profiles under high pressure and low temperature [19]. A higher proportion of unsaturated fatty acids would help maintain favourable fluidity and viscosity of biological membranes under high pressures. It is noteworthy that piezophiles present higher metabolic activities under high pressures. Incubating hydrothermal sediments from Guaymas basin under variable pressures, observed higher sulfate reduction rates are higher at 45 MPa than at 1 Mpa in the 75–100°C temperature range [20].

2.5 Porosity

As depth increases, burial of sediments and diagenesis limit space availibility for microorganisms due to the decrease of porosity. So geometrical constraints and mechanical interactions must be considered as well. Small pores restrict bacterial movement and activity, limit nutrient transport and slow the rate of bacterial division. Sediments compaction also leads to spatial isolation due to a lack of pore connectivity, and the decrease of pore-throat size. This implies that all cells in a pore are lineal descendants of a bacterium that became entombed at the time of geologic deposition [21].

3. Tools for analyzing microbial communities

Microbial communities can be analyzed by two major ways, cultivation and culture-independent nucleic acids analyses.

3.1 Culture dependent methods

Bacterial cultures lead to enriched cultures and isolation of pure strains. It is clear that pure cultures do not exist in nature, and that cultivable bacteria only represent a minor fraction of total prokaryotic biomass [22]. In most cases, cultivation-based methods fail to detect the most abundant members of microbial communities in situ. Nevertheless, cultivation of bacterial strains is necessary to access the real physiological mechanisms, and so remain crucial. This is particularly true within the oil field water where there is less biodiversity, as a result of physico-chemical constraints. A most-probable-number

(MPN) method is generally used for the quantitative assessment of microbial communities, with a defined medium containing carbon compounds or not in combination with different electron acceptors targeting different physiological groups (e.g. sulfate-reducers, nitrate-reducers, FeIII-reducers, methanogens and even microaerophiles when present…) or different fermentescible sugars targeting heterotrophic bacteria.

3.2 Culture-independent methods

In parallel, nucleic acids (DNA and RNA) extraction and analyses can be performed. DNAs are large and stable molecules that contain the bacterial genome, thus the potentialities of the microorganisms; RNAs are small and short-lived molecules, formed from DNA, which can be used as a proxy for microorganism activity. Nucleic acids sequences and concentration from deep subsurface can give great informations about microbial ecology and physiology of the microbes (e.g. in situ activities).

A large and diverse suite of protocols has been published on nucleic acids extractions. Two principal approaches exist with their advantages and their drawbacks. The principal approach used consists in an in situ lysis of microbial cells within the environmental matrix (soil, sediment, rock) followed by separation of the nucleic acids from matrix components and cell debris. These methods lead to higher nucleic acids yields and are less time-consuming. But directly-extracted nucleic acids often contain considerable amounts of co-extracted susbtances (such as humic acids in the case of soils) that interfere with subsequent molecular analysis. Furthermore, in certain cases (in particular soils), a proportion of extracted DNA might originate from non-bacterial sources or from free DNA.

Once nucleic acids are extracted and purified, they can be analyzed by way of molecular methods. The principal targeted gene is the 16S rRNA gene. This gene is present in all Prokaryotes. Its product (16S rRNA) is part of the ribosomes which are required by all organisms to synthetize new proteins. It is sufficiently long (1500 bp) to be used as a document for evolutionary history and there is no evidence for horizontal transfer of this gene. As 16S rRNA genes consist of several sequences domains that have evolved at different rates (due to functional constraints), unambiguous alignment of homologous sequences (to define primers and probes) and identification of taxon-specific signatures are possible. However, when considering a single 16S rRNA gene fingerprint band or a 16S rRNA gene partial sequence (see below), the phylogenetic level which this band or sequence represents is not absolutely sure. Many researchers are using the term "Operational Taxonomic Unit" (OTU) to define each band profile or groups of highly related sequences. The use of arbitrarly defined OTU is acceptable from a taxonomic point of view since the definition of bacterial species is itself somewhat arbitrary, as no single marker is also sufficient and necessary [23].

PCR (polymerase chain reaction) is the first step in phylogenetic analysis of microbial communities and for many molecular ecological approaches.

PCR allows the selective amplification of small amounts of DNA extracted from natural samples. The oligonucleotide primers are designed to hybridize regions of DNA flanking the desired gene sequence. Primer selection will depend on the particular application and whether 16S rRNA genes from a wide range of organisms or specific groups are targeted. Repetitive cycles of temperature changes result in an exponential increase in the DNA fragment of interest. Some care must be taken into account when using PCR-based methods. PCR pitfalls are point mutation or deletions, the formation of chimeric PCR products, differential amplification leading to a biased picture of the community [24, 25].

At the end of PCR, the mixture of 16S rRNA genes of comparable lengths can be analyzed several approaches: construction of clone libraries, DNA fingerprinting, and nowadays, high throughput sequencing. DNA sequences can be read and provide information on taxonomic affiliation of microbes as on metabolic capacities.

4. Microbial heterogeneities in subsurface

Improving understanding of the spatial and temporal distribution of microbial organisms and activities is critical due to the difficulty and high cost of obtaining large numbers of well-preserved subsurface samples. The microbiological properties appear to be spatially correlated to geologic, hydrologic and geochemical properties [14, 21]. Subsurface environment often exhibits a pronounced spatial heterogeneity in physical properties as a result of stratified structures and the actions of hydrologic and geochemical processes over geologic time periods. Movements of fluids are strongly controlled by geological setting. As a result, subsurface environments may possess strong contrasts including variations in redox conditions, moisture and nutrients fluxes. Thus, there is potential high heterogeinity in microbiological processes in subsurface. The scale that should be investigated is a function of the scientific question or applied problem under investigation. Studying bioremediation design and engineering, microbial alteration of physical properties and dynamics of microbial colonization should concern strata, laminae and pore/pore network, respectively. Zones of enhanced activity or populations may exist at interfaces of strata where electron donors and acceptors mix [26]. At 90 m depth in a margin site, Pacific Ocean, Prokaryotes had greater activity than in near-surface sediments: this zone corresponded to the sulfate-methane transition, where both sulfate reduction and methane production occurred [26]. The low amounts of methane measured were attributed to anaerobic methane oxidation by a consortium of methane oxidizers (Methanosarcinales) and sulfate reducers (δ-Proteobacteria) [27]. Results indicated that subsurface Prokaryotes were highly active, presented changing diversity associated with interfaces, and that they were active over geological timescales. Geostatistics focuses on the spatial patterns of data, providing tools for characterizing spatial distributions and estimations of variables at unsampled locations. The spatial continuity on which the geostatical approach is based results from the inherent continuity of the processes that control earth sciences phenomena. For example, the geochemical and hydrogeological processes controlling microorganisms distribution in subsurface do not vary randomly, but exhibit strong spatial continuity. Thus, geostatistical models (variogram analysis, cross-correlations...) could increase the ability to model and predict the distributions of microbial properties and the responses of microbial communities to environmental perturbations.

In contrast to studies of surface microbial communities, studies of subsurface environment have not adequately characterized microbial community composition and diversity because of the cost and difficulty of obtaining a large number of samples. It is not clear whether the subsurface microbial communities are closely tied to the surface soil communities, or belong to an independent environment. Subsurface communities are isolated from each other to a greater degree than soil communities. In surface soils, wind transport and flooding can move and mix soils over significant distances, which can not be done in subsurface. Therefore, community differences between sites separated by relatively small distances are potentially greater for subsurface communities. In subsurface, there is the possibility of the existence of some microbial communities linked to the process of sediment deposition, million years ago rather than to their surface counterpart [28].

5. Conclusion

With the development of new high-throughput sequencing technologies, since 2005, pictures of deep subsurface communities from great quality are now being provided, with some obvious results: it seems for example that in certain environments, the communities are dominated by rare taxons [29]. Physiological studies reveal that famine could be the most common lifestyle in deep subsurface [30], communities being prone to reactivation when amended [31].

References

[1] H.W. Jannasch, K. Eimhjell, C.O. Wirsen, A. Farmanfa, Science **171** (1971).
[2] R.J. Parkes, B.A. Cragg, S.J. Bale, J.M. Getliff, K. Goodman, P.A. Rochelle, J.C. Fry, A.J. Weightman, S.M. Harvey, Nature **371** (1994).
[3] J.C. Fry, R.J. Parkes, B.A. Cragg, A.J. Weightman, G. Webster, FEMS Microbiol. Ecol. 66 (2008).
[4] E.V. Pikuta, R.B. Hoover, J. Tang, Crit. Rev. Microbiol. **33** (2007).
[5] J.P. Amend, E.L. Shock, FEMS Microbiol. Rev. **25** (2001).
[6] W.B. Whitman, D.C. Coleman, W.J. Wiebe, Proc. Natl. Acad. Sci. U.S.A. **95** (1998).
[7] I.M. Head, D.M. Jones, S.R. Larter, Nature **426** (2003).
[8] L.H. Lin, G.F. Slater, B. Sherwood Lollar, G. Lacrampe-Couloume, T.C. Onstott Geochim. Cosmochim. Acta **69** (2005).
[9] T.M. Hoehler, M.J. Alperin, D.B. Albert, C.S. Martens Geochim. Cosmochim. Acta **62** (1998).
[10] J. Debussy, M. Pagel, J.M. Beny, H. Christensen, B. Hickel, C. Kosztolanyi, B. Poty, Geochim. Cosmochim. Acta **52** (1988).
[11] B. Sherwood Lollar, T.D. Westgate, J.A. Ward, G.F. Slater, G. Lacrampe-Couloume, Nature **416** (2002).
[12] A.I. Rushdi, B.R.T. Simoneit, Orig. Life Evol. Biosph. **31** (2001).
[13] G.D. Cody, N. Boctor, J. Blank, J. Brandes, N. Boctor, T. Filley, R. Hazen, H.J. Yoder, Orig. Life Evol. Biosph. **30** (2000).
[14] F. Inagaki, M.M. Kuypers, et al., Proc. Natl. Acad. Sci. U.S.A. **103** (2006).
[15] A. Oren, Microbiol. Mol. Biol. Rev. **63** (1999).
[16] F. Abe, C. Kato, K. Horikoshi, Trends Microbiol. **7** (1999).
[17] A.A. Yayanos, E.C. Pollard, Biophys. J. **9** (1969).
[18] M. Gross, K. Lehle, R. Jaenicke, K.H. Nierhaus, Eur. J. Biochem. **218** (1993).
[19] E.F. DeLong, A.A. Yayanos, Science **228** (1985).
[20] J. Kallmeyer, A. Boetius, Appl. Environ. Microbiol. **70** (2004).
[21] V. Rebata-Landa, C.J. Santamarina, Geophys. Geochemi. Geophysics. Geosyst. **7** (2006).
[22] R.I. Amann, W. Ludwig, K.H. Schleifer, Microbiol. Rev. **59** (1995).
[23] R. Rosselló-Móra, R. Amann, Syst. Appl. Microbiol. **38** (2015).
[24] A.L. Reysenbach, L.J. Giver, G.S. Wickham, N.R. Pace, AEM **58** (1992).
[25] M.C. Hansen, T. Tolker-Nielsen, M. Givskov, S. Molin, FEMS Microbiol. Ecol. **26** (1998).
[26] R.J. Parkes, G. Webster, B.A. Cragg, A.J. Weightman, C.J. Newberry, T.G. Ferdelman, J. Kallmeyer, B.B. Jorgensen, I.W. Aiello, J.C. Fry, Nature **436** (2005).
[27] A. Boetius, K. Ravenschlag, C.J. Schubert, D. Rickert, F. Widdel, A. Gieseke, R. Amann, B.B. Jørgensen, U. Witte, O. Pfannkuche, Nature **5** (2000).

[28] G. Gales, N. Tsesmetzis, I. Neria, D. Alazard, S. Coulon, B.P. Lomans, D. Morin, B. Ollivier, J. Borgomano, C. Joulian, Sci. Rep. **11** (2016).
[29] S. De Mandal, S. Zothansanga, A.K. Panda, S.S. Bisht, N. Senthil Kumar, Environ. Sci. Pollut. Res. **23** (2016).
[30] T.M. Hoehler, B.B. Jørgensen, Nat. Rev. Microbiol. **11** (2013).
[31] T.T. Fida, C. Chen, G. Okpala, G. Voordouw, Appl. Environ. Microbiol. **6** (2016).

R&D of a SIMPLE bubble chamber for dark matter searches

M. Felizardo[a], T.A. Girard, I. Lazaro, A.C. Fernandes, T. Morlat, J.G. Marques, and A. Kling

C^{2}TN, Centro de Ciências e Tecnologias Nucleares, Instituto Superior Técnico, Universidade de Lisboa, E.N. 10 ao km 139,7, 2695-066 Bobadela LRS, Portugal

Abstract. SIMPLE (Superheated Instrument for Massive ParticLe Experiments) intends the realization and full field tests of a 25 kg, low background, low frequency-instrumented BC (Bubble Chamber), with data from the BC testing and a small array demonstration providing new physics results. We report on the development of a 1 kg freon BC prototype, including chamber recompression system design and testing with initial acoustic detection of bubble formation. Realization of full field tests leading to a 25 kg, overall 10^{-6} recoil evt/kgF/d background, low frequency-instrumented BC, incorporating several innovations designed, are intended to serve for an anticipated 2500 kgd exposure, resulting in new exclusion contours with sensitivities of 2×10^{-8} and 5×10^{-5} pb at 35 GeV in the SI and SD sector of the WIMP phase space, respectively.

1. Introduction

The direct detection of astroparticle dark matter has returned to the spotlight of forefront physics. Its existence is now well-established by indirect means, to include a wealth of observational data but, despite the progress made in the direct search effort, over 10 orders of magnitude in sensitivity since 1984, no candidates have appeared in searches covering a range from about 10 GeV to 1 TeV. Such astroparticle candidates are the leading candidates for dark matter particles, fitting nicely into the standard model of particle physics. Among the several technologies in current use is the bubble chamber (BC), based on the phase transition of a superheated liquid to its gas phase by energy deposition in the liquid, which must fulfil two thermodynamic conditions, which together render the detector effectively insensitive $>10^{10}$ to the majority of traditional particle backgrounds (γ's, β's, cosmic μ's) which plague the other more conventional search detectors. Additional advantages over conventional techniques include a relative inexpensiveness, a relative high efficiency in effective/raw exposure ($\sim$80% vs. 40–50%), a simpler data acquisition/analyses, an ability to achieve low recoil energies, and reduction of the possible WIMP phase space via easy substitution of different superheated liquids. In short, the technique ranks among the most competitive dark matter search approaches. A BC requires a recompression system: this is based on a simple piston configuration to prevent N_2 recompression gas from refrigerant/gel contact because of solubility, and for robustness. Its basic response to a nucleation event is a sensed pressure increase which provokes an immediate piston drop to increase the pressure to 10 bar (at which the gas is recompressed to a metastable liquid state), followed by a pressure

[a] Corresponding author: felizardo@ctn.tecnico.ulisboa.pt

decrease to the operational pressure for the superheated state (2 bar in the case of C_2ClF_5); the entire recompression cycle requires ~200 s. Any acoustic signal related to gas bubble nucleation and growth in the liquid halocarbon must appear in coincidence with the pressure increase resulting from the same process. The shock wave produced by the bubble nucleation and expansion can be recorded by an acoustic sensor and converted into an electrical pulse. This pulse tends to be a damped sinusoid with a typical duration of several milliseconds. The projected BC sensitivity for dark matter searches demands an intrinsic background contribution of $\leq 10^{-6}$ events/kgF/d.

2. Past SIMPLE studies and field tests

Phase II of SIMPLE searched for astroparticle dark matter using superheated liquid C_2ClF_5 droplet detectors. Each droplet generally required an energy deposition with linear energy transfer (LET) $\geq$150 keV/μm for a liquid-to-gas phase transition, providing an intrinsic rejection against minimum ionizing particles of order 10^{-10}, and reducing the backgrounds to primarily α and neutron-induced recoil events. The droplet phase transition generated a millimetric-sized gas bubble that was recorded by acoustic means. We described the SIMPLE detectors, their acoustic instrumentation, and the characterizations, signal analysis and data selection, which yielded a particle-induced, "true nucleation" event detection efficiency of better than 97% at a 95% C.L. The recoil-α event discrimination, determined using detectors first irradiated with neutrons and then doped with alpha emitters, provided a recoil identification of better than 99%. The science measurements, comprising two shielded arrays of fifteen detectors each and a total exposure of 18.24 kgd, were thoroughly detailed. These measurements provided new contour minima of $\sigma_p = 4.3 \times 10^{-3}$ pb at 35 GeV/c^2 in the spin-dependent sector of astroparticle dark matter–proton interactions and $\sigma_N = 3.6 \times 10^{-6}$ pb at 35 GeV/c^2 in the spin-independent sector. These results were examined with respect to the fluorine spin and halo parameters used in the previous data analysis.

3. Bubble chamber developments

The principle challenges are a low cost 25 L BC capable of withstanding repeated 16 bar overpressures while recording recoil events with an overall background rate of $\sim 10^{-6}$ evts/kgF/d, and α-recoil discrimination ability with better than 99% efficiency. The essential engineering R&D, must be working hand-in-hand with the radioactive contaminants of the materials otherwise suitable for chamber construction. An example of one of the 4 prototype chambers and 2 recompression systems is seen in Fig. 1. The recompression system (Fig. 1 (left)) simultaneously controls two separate BCs (~20 L volume each), with a 2 s response and 164 s overall recycling times, with both the time and pressure threshold adjustable. Larger chamber volumes require a larger substitution of the current piston.

The chambers of Fig. 1 (right) consisted of a commercially-available home water filters capable of withstanding a repeated 10 bar over-pressuring, with an estimated overall background contribution of $\sim 10^{-2}$ recoil evts/kgF/d from radio-assays. It possessed an internal 1 cm thick, transparent, purified gel sheathing to reduce homogeneous nucleation events [1]. The proposed chamber design concept differs from Fig. 1b, primarily by the extension of the sheathing to $\geq$ 10 cm because of the U/Th content of the commonly-available, inexpensive plastic containments assayed at $\sim 10^{-2}$ recoil evts/kgF/d. It consists of a solid block of purified gel in which a cylindrical liquid containment hole is bored such that the bore hole is surrounded by a minimum 10 cm of gel 4π. The gel itself is contained by a low background material such as acrylic plastic, with the contribution of its U/Th contaminants

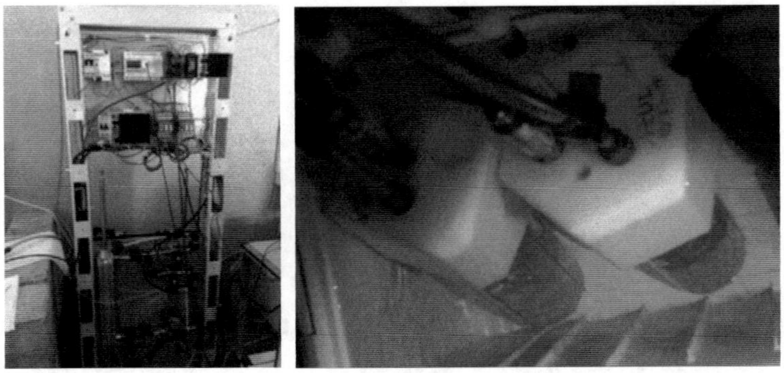

Figure 1. Left: piston-based recompression system, right: 1 kg C_2ClF_5 BCs.

moderated by $\sim 10\times$ by the displacement. A chamber pressure increase is easily triggered by a spontaneous nucleation event originating in imperfections in the containment surface: use of the previously-tested gel sheathing, which smoothes this interface, results in a $\sim 20\times$ increase in time between recompressions [2].

BC pressure control is assured via a piston pushing or releasing high purity glycerin as necessary into the container. The halocarbons used as active liquids in our tests are immiscible in glycerin and denser, thus falling to the gel sheeted bottom of the container, leaving the glycerin to provide a smooth interface between halocarbon and detector cap. A set of electrovalves controls the system. The electrovalves are controlled via the relay outputs of a Crouzet M3 Millenium logic controller, which continuously monitors the BC pressure signal from a pressure transducer assembled on the container. The BC temperature is measured by means of a thermocouple type k, inserted in the BC glycerin through a feedthrough in the cap.

4. Backgrounds & simulations

The purified gel sheathing provides a further shielding to both the containment U/Th decays radiations and diffusion of external α-producing radon into the chamber (which are problematic because of (α, n) reactions with the materials). The projected WIMP search sensitivity demands an intrinsic background contribution to the measurements of $\leq 10^{-6}$ events/kgF/d. Previous materials' contributions, was at the level of 10^{-4}–10^{-5} evt/kgF/d [3]. This level is to be reduced via a series of new materials selections based on (a) materials characteristics and (b) inherent U/Th composition. The performed radioassays of materials samples are executed by using a combination of chemical analyses of compositions, ion beam analysis of the hydrogen content of the materials (elastic recoil detection and nuclear reaction analysis + elastic backscattering, respectively), γ/α spectroscopy, and comparative neutron activation analysis. Materials requiring increased sensitivity levels, particularly in Th.

The radio-assay results are used in existing MCNP transport simulations [3] of the expected detector response with modified programs already developed. Since neutron yield and spectrum in a particular material depend not only on its radio-purity but also on its material composition and formation (e.g. the U/Th decay chains in steel are not in secular equilibrium because of the fabrication process), changing of a material (for instance, using plastic instead of PTFE or aluminium instead of stainless steel) to reduce the background rate will be investigated. Radio-assay of various plastics identified materials with a factor 10 less background; the internal sheathing can be increased to distance the containment (as well as a previously-developed spatial-resolving system implemented [4]). Recoil simulations are also

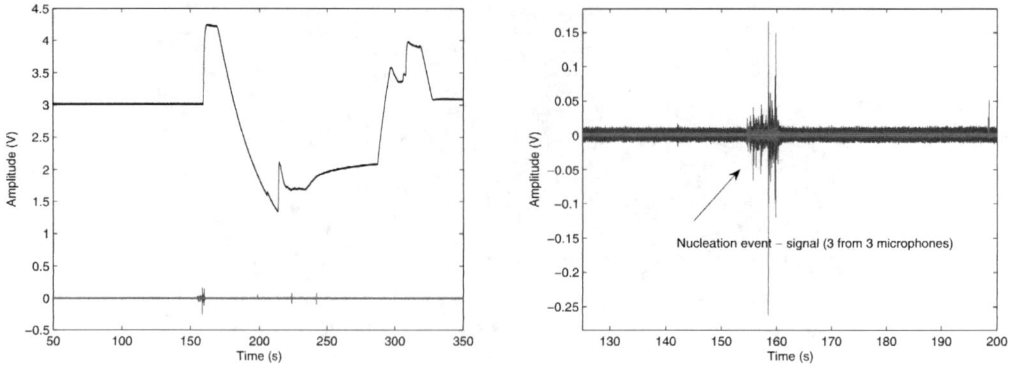

Figure 2. Left: BC pressure signal (top) and signal from the three microphones coupled to the PET container (bottom) during a nucleation/recompression cycle of a C_2ClF_5 filled BC at 9°C, where the halocarbon was allowed to boil for some time without triggering the recompression system, right: zoomed nucleation events captured by the 3 low frequency microphones.

performed for calibration sources/configurations, to ensure: (i) the on-target source event rate exceeds background at least 10x, and (ii) a low rate of pile-up events.

5. Nucleation event signal detection

As recently demonstrated the sound of a gas bubble nucleation in a bubble chamber shows a broad emission up to a frequency of 250 kHz. For this reason, two different types of microphones for the SIMPLE BC prototype have been used: (a) An externally-polarized condenser microphone (CM16/CMPA40-5V from Avisoft-Bioacoustics), with a flat response over the 10–150 kHz frequency range (high frequency microphone); (b) A PANASONIC (omnidirectional) high quality electret microphone (MCE-200), with a frequency range of 20 Hz–16 kHz (3 dB), with a signal-to-noise ratio (SNR) of 58 dB and a sensitivity of 7.9 mV/Pa at 1 kHz, already used with success for acoustic nucleation event detection in SDDs [5]. Several bubble chambers were assembled and deployed for testing at the Lisbon SIMPLE laboratory; one was filled with with ~800 g of C_2ClF_5 and deployed at 9°C in temperature (kept constant in a water bath) and 2.0 to 3.0 bar pressure (absolute). Figure 2 shows the preliminary results of experiment using the C_2ClF_5-filled BC at 9°C. One can see the BC pressure signal –top- and microphone signals –bottom- during an event/recompression cycle in the C_2ClF_5 filled BC at 9°C and 3 bar pressure. The signals were digitized with the NI PCI-6251 DAQ, at an acquisition rate of 300 kS/s.

Preliminary frequency spectrum of the acoustic signal identified as resulting from a nucleation event. The recompression system allowed for a reliable pressure trigger from BC pressure differences (Δp) resulting from gas bubble nucleation as low as Δp=100 mbar, without spurious recompression cycles. The time between the observed pressure increase and the maximum recompression pressure being reached depends on the N_2 pressure and the length of the tubing connecting the piston and the BC. Typically, for a 5 m long, 10 mm diameter tube, it is around 6 seconds. Decompression times can be regulated by means of a fine metering valve, and were typically set in the 30–60 sec interval. The noise level throughout the trials was 6.27 ± 1.37 mV.

Analyses of the frequency microphone signals placed on the container external wall around this pressure/time stamp indicate the presence of an acoustic pulse with shape similar to a true nucleation event from a SDD. Figure 3 shows the low frequency microphones

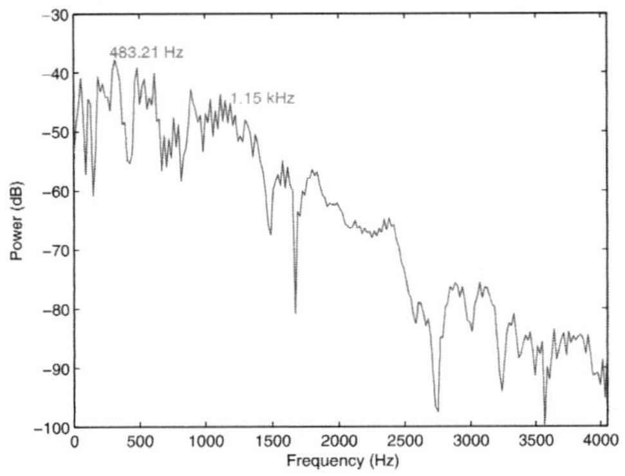

Figure 3. Frequency spectrum of the nucleation events seen in Fig. 2.

response on the container walls of the events in Fig. 2. There is a clear low frequency component with several harmonics perfectly distinguishable from other external sources of noise. Data was analysed as in the previous SIMPLE phase [5].

These preliminary results are of great importance for future BC dark matter searches since they can permit with the intrinsic background expected a clear discrimination of acoustic backgrounds from particle-induced signals with low frequency instrumentation, which has been well-demonstrated, as well as discriminations between neutron- and alpha-induced events with previous SIMPLE detectors [6–9].

6. Future performance tests – response studies and field tests

Following successful response testing of each BC, it will be taken to the LSBB [10] for full field, low background testing at 1500 mwe underground. The measurement site (GESA) is $60\,m^3$, containing $\sim$8 ton of water shielding surrounding a HDPE + wood platform supporting a temperature-controlling 400 L water-bath. New background event simulations are made for neutrons originating from U/Th contaminations and (α,n) reactions of the environment, as well as new shielding components [4]. Previous SIMPLE results have shown that the expected GESA background rate excluding its water-shielding is at the level of 10^{-5} evt/kgF/d. The water-bath and shield water, originally simply tap water with a background contribution of $\sim 10^{-3}$ recoil evts/kgF/d, is to be resin-purified which reduces the heavy metals content to 2×10^{-4} g/g, from which a factor 100 decrease in the current water U/Th content to yield $\sim 10^{-5}$ evt/kgF/d is estimated [3].

7. Impacts

Successful completion of the response measurements alone results in new "zero-background" 100–500 kgd exposures providing new results competitive with the current direct measurements, and comparable with both indirect and LHC results; successful project completion also results in a demonstrated new tool in the physics detectors "toolbox" for rare event search activities, which can be rapidly upgraded to larger, multi-ton arrays for deployment. With a 2-BC field test measurements, the results may contribute immediately to the question of astroparticle dark matter. The projected "zero-background" limits are shown in Fig. 4, indicating sensitivities of order 10^{-4} (SD), 10^{-7} (SI) pb, providing further evidence for/against an astroparticle explanation for dark matter.

DOI: 10.1051/e3sconf/20161203002

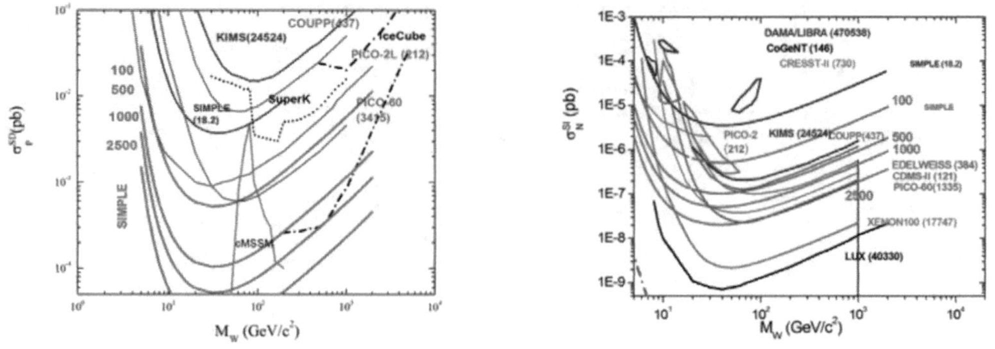

Figure 4. Left: background-free S_pD results with C_2ClF_5 at 6 keV recoil detection threshold; right: same for background-free SI results.

8. Conclusions

Rare event searches are by nature an expensive technology arena, in terms of both their search-dictated large mass and exposure requirements. A fully functional SIMPLE prototype bubble chamber and its command and control system were developed that mitigates the initial assumptions of costs. Successful project accomplishment opens a new pathway to resolution of the question of Astroparticle dark matter existence. The anticipated results should equal or surpass the leading results. The BC technology (thermodynamic response, acoustic instrumentation) differs from others in its mechanism, opening new lines of industrial involvement.

This work is supported by C^2TN/IST and by grants PTDC/FIS/115733/2009, PTDC/FIS/121130/2010, PTDC/EEI-ELC/2468/2014 and IF/00628/2012/CP0171/CT0008. C^2TN/IST authors gratefully acknowledge the FCT support through the UID/Multi/04349/2013 project. The activity of M. Felizardo is supported by FCT grant SFRH/BPD/94028/2013.

References

[1] T.A. Girard, et al.: Proc. (iDUST -2014), Apt, France, May 5–7 (2014).
[2] J. Puibasset, PhD Thesis, Univ. Paris, (unpublished) (2000).
[3] A.C. Fernandes, et al., *Nucl. Instr. & Meth.* A **623** (2010) 960.
[4] M. Felizardo, et al., *Nucl. Instr. & Meth.* A **599** (2009) 93.
[5] M. Felizardo, et. al., *Nucl. Instr. Meth.* A **589** (2008) 72.
[6] M. Felizardo, et al., *Phys. Rev.* D **89** (2014) 072013.
[7] M. Felizardo, et. al., *Phys. Rev. Lett.* **105** (2010) 211301.
[8] M. Felizardo, et. al., *Phys. Rev. Lett.* **108** (2012) 201302.
[9] A.C. Fernandes, et al., *Astroparticle Physics* **76** (2016) 48–60.
[10] LSBB `http://www.lsbb.eu`

Microbiology of shallow subsurface aquifer and carbonate rocks studied by culture-dependent and culture-independent methods. Preliminary results on an underground laboratory, the LSBB, Rustrel, France

Grégoire Galès[1,2,a], Gaël Erauso[2], Jean-Luc Cayol[2], David Navarro[3], Daniel Boyer[4], and Gilles Micolau[5]

[1] Aix-Marseille Université, CEREGE, Centre St Charles, Case 67, 3 place Victor Hugo, 13331 Marseille, France
[2] Aix-Marseille Université, Université du Sud Toulon-Var, CNRS/INSU, IRD, MIO, UM 110, 13288 Marseille Cedex 09, France
[3] INRA, UMR 1163 Biodiversité et Biotechnologie Fongiques, Marseille, France
[4] LSBB, University of Nice, University of Avignon, CNRS, Aix-Marseille University, La grande combe, 84000 Rustrel, France
[5] UMR EMMAH Environnement Méditerranéen et Modélisation des Agro-Hydrosystèmes, INRA/Université d'Avignon et des Pays de Vaucluse, 33 rue Louis Pasteur, 84000 Avignon, France

Abstract. We investigated the microbiology of a shallow subsurface site, the LSBB, located near Avignon. This site lies in carbonate rocks, belonging to the Urgonian facies. Rock, concrete and water samples were collected and directly transferred to the laboratory. Studies of microorganisms as pure cultures are the only way to get their real physiological properties. Nevertheless, microbiologists cannot cultivate and isolate the majority of microorganisms for several reasons, one being our lack of understanding of their minimal needs. Molecular studies, e.g. extraction and sequencing of the total nucleic acids present in an environment provide phylogenetic and metabolic information on uncultivated microorganisms. We performed aerobic and anaerobic culture with various electron acceptors and donors, searching for heterotrophic, methanogenic, sulphate- nitrate- and FeIII-reducing Prokaryotes. We also performed DNA extractions and PCR amplification of ribosomal RNA genes, to test if our protocols were adapted to this environment. Our results show that the LSBB galleries are colonized by a low diversity microbiote, with a strong influence of anthropogenic activities. Further studies will link the microorganisms biodiversity and the petrophysic properties of rocks.

1. Introduction

It became apparent during the last decades that life could occur deeper than previously presumed in the underground or subsurface [1]. Several deep habitats, including geothermal springs, oceanic crust, gold mines, oil reservoirs... were colonized by a large diversity of Prokaryotes (microorganisms which live without a cell nucleus, e.g. *Archaea* and *Bacteria*).

[a] Corresponding author: gregoire.gales@univ-amu.fr

 DOI: 10.1051/e3sconf/20161205002

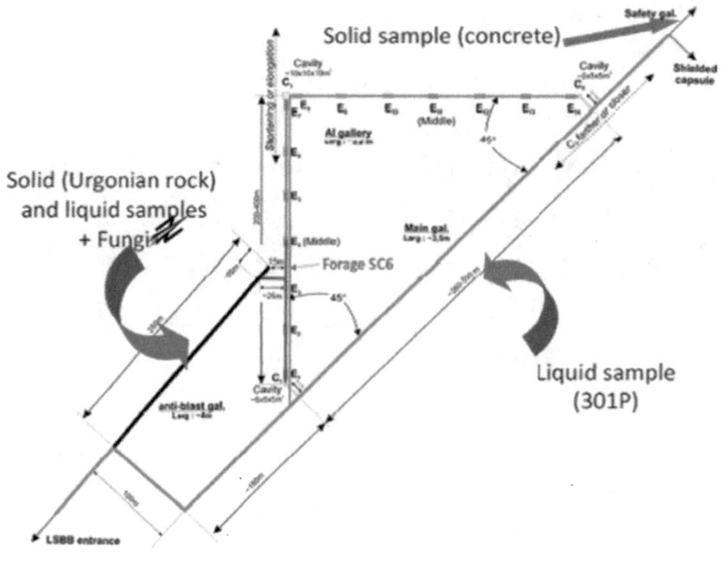

Figure 1. Localization of the sampling area.

Carbon sequestration in subsurface could be one of the ways to reduce the effect of anthropogenic carbon dioxide emissions. Bacterial-induced carbonate precipitation allows the filling of fracture and the optimization of injected quantities, thus being a concern on safety and efficiency [2]. To understand if indigenous subsurface microflora could affect carbon dioxide sequestration, we studied the microbiology of a shallow subsurface site, the LSBB, located near Avignon. This site lies in carbonate rocks, belonging to the Urgonian facies. We present here some preliminary results.

2. Material and methods

2.1 Sampling

Drilling deep geological formations through hundred meters generates costs higher than further microbiological analyses. Many subsurface studies benefit from windows open into earth crust, like mines and drillings for oil-producing wells. But drilling and building oil- or water-wells have considerable impact on the ecosystems. Inside the well tubing, physico-chemical conditions are far from those prevailing in subsurface and microbial biofilm adhere to the well pipes. Therefore, microbial communities sampled from oil- and water-wells do not obligately reflect their subsurface counterpart.

We chose to sample rock, concrete, biological (a Fungi) and water samples. Samples location is indicated on Fig. 1. These samples were directly transferred to the laboratory for further characterization.

2.2 DNA extractions and PCR

Rock samples were crushed in a mortar under anaerobic and sterile conditions. Crushed rock samples (0.3 g) or water (500 μl) or harvested cells from pure cultures (500 μl) were suspended in 1 mL TE-Na-1X lysis buffer (100 mM Tris, 50 mM EDTA, 100 mM NaCl, 1% SDS (w/v), 1% (w/v) Sarkosyl, pH 8.0). Cells were disrupted by bead beating using

DOI: 10.1051/e3sconf/20161205002

Lysing Matrix E (MP, Biomedicals). After digestion with proteinase K (1 mg.ml^{-1} final) for 1 h at 55°C, DNA was extracted twice with a volume of phenol/chloroform/isoamyl alcohol (25/24/1; pH 8.0), then with a volume of chloroform. Aqueous and organic phases were separated by centrifugation at 13,000 g for 15 min at 4°C. DNA was precipitated with 0.7 volume of isopropanol, pelleted by centrifugation at maximum speed for 15 min, then air-dried and finally resuspended in 30 μL of TE 1X (10 mM Tris,.HCl, 2 mM EDTA, pH 8.0).

The 16S rRNA gene was amplified by PCR using the universal primers Rd1 (5'-AAGCGGTTGCCGCCGACCGACC-3') and Fd1 (5'-AGAGTTTGATCCTGGCTCAG-3'). 16S amplicons were then stored at – 80°C before analyses (work in progress). The Fungi was studied by the amplification of the ITS6 region.

2.3 Culture of microorganisms

A basal medium was defined, containing (l^{-1}): 2 g NaCl, 0,5 g K_2HPO_4, 0,5 g KH_2PO_4 and 0,5 g yeast extract. The pH was adjusted to 8.5 (corresponding to the pH measured from the rocks and water samples) with 10 M KOH. To target some physiological prokaryotic groups, we made aerobic (targeting aerobic microorganisms) and anaerobic cultures. To obtain anaerobiosis, the medium was boiled under a stream of O_2-free N_2 gas and cooled at room temperature. Aliquots of 5 ml were dispensed into Hungate tubes, degassed under N_2/CO_2 (80 : 20, v/v) and subsequently sterilized by autoclaving at 120°C for 20 min. After autoclaving, the gas phase was changed for a H_2/CO_2 (80 : 20, v/v) mix. In anaerobic cultures, we provided when needed nitrate and succinate (10 mM each, targeting nitrate-reducing microorganisms), $FeCl_3$ and citrate (10 mM each, targeting Iron-reducing microorganisms), lactate and sulphate (10 mM each, targeting sulphate-reducing microorganisms) and methanol and formiate (10 mM each, targeting methanogenic *Archaea*). After microscopic observations of liquid cultures, we inoculated solid media (Petri dishes or Roll-tubes) to obtain pure colonies. Cultures were performed at ambient temperature.

DNA from pure colonies was extracted as explained before, 16S rRNA gene was also amplified and sequenced. 16S rRNA genes of pure cultures were determined and analyzed using the RDP database [3, 4]. The sequences were compared with available sequences in GenBank, using a BLAST search. For a sequence showing only 97% similarity with that of *Pectobacterium aroidearum*, we built a phylogenetic tree whose topology was determined using the neighbour-joining method (Fig. 2). Support for internal nodes was assessed by bootstrap analysis (1500 replicates) [5]. These analyses were conducted in MEGA6 [6].

3. Results

3.1 DNA extractions and PCR

Among the factors that limit the ability to extract nucleic acids from carbonate reservoirs is the low number of bacterial cells generally associated with nutrient depleted environments and the chemical nature of the rock. Calcite binds DNA with a high affinity, due to the strong chelating affinity of calcium ions for the phosphate backbone of the DNA molecule. Nevertheless, DNA extractions from rock and water samples gave positive results, as we recovered DNA prone to PCR amplification. Further analyses will be done on these DNA, namely sequencing of the 16S rRNA genes, to get a picture of the micoorganisms inhabiting the LSBB galleries.

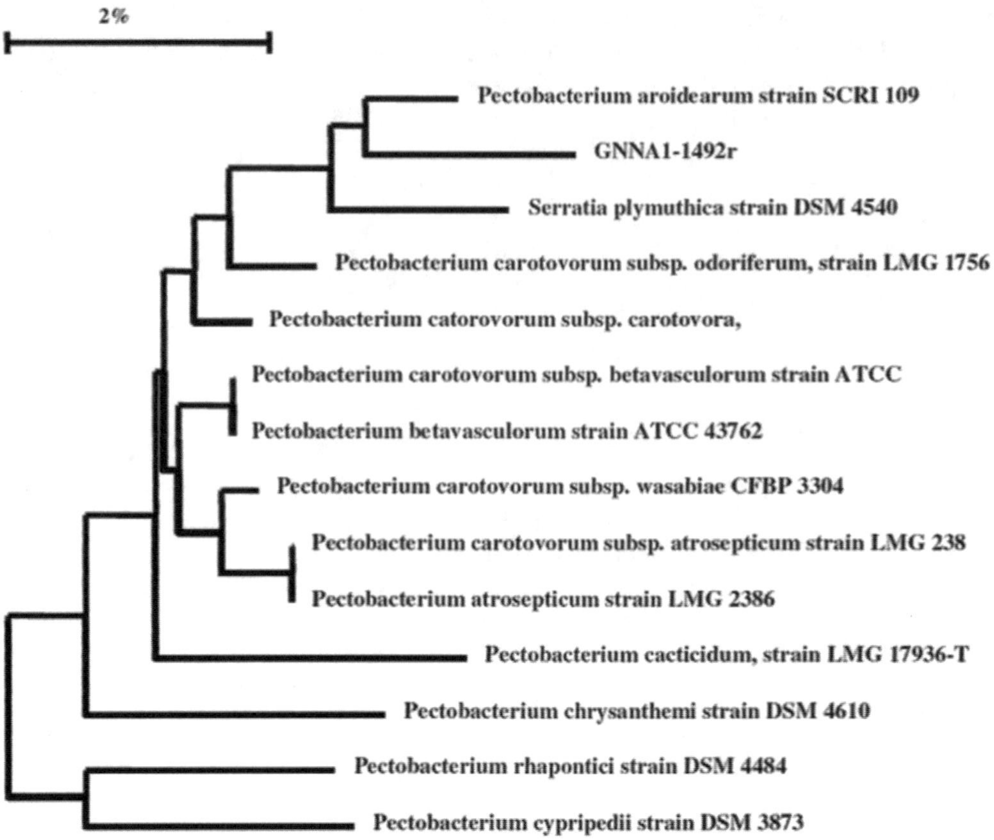

Figure 2. Neighbour-joining phylogenetic tree of 16S rRNA gene sequence showing the relationships between strain GNNA1 and related type strains amongst Proteobacteriales.

Table 1. Functional diversity of microorganisms recovered from the LSBB. + means presence and isolation of the microorganisms, +* means presence but not isolation (work in progress) and Ø means absence.

Sample location	**Fermentative or respiratory**	**Nitrate-reducing**	**Sulfate-reducing**	**FeIII-reducing**	**Methano-genic**
Anti-Blast Gallery **Solid**	+	+	Ø	+*	+*
Anti-Blast Gallery **Liquid**	+	Ø	Ø	Ø	Ø
Main Gallery **Liquid**	+	Ø	Ø	Ø	+*
Safety Gallery **Solid**	+	Ø	Ø	Ø	Ø

3.2 Isolation of microorganisms

Several aerobic microorganisms were isolated from solid and liquid samples. Functional diversity of microorganisms is given Table 1.

Microorganisms isolated belonged to genera like *Bacillus, Micrococcus, Acidovorax, Herbaspirillum* and *Stenotrophomonas*. They were closely affiliated (more than 99%) to known and described species. We found in nitrate-reducing conditions a strain, GNNA1, whose affiliation is not sure, closely related to *Pectobacterium aroidearum*. We built a

 DOI: 10.1051/e3sconf/20161205002

phylogenetic tree with 1000 bases (Fig. 2), showing that it could be a new species within the *Pectobacterium* genus. New sequencing of the 16S rRNA gene will be performed, to get the entire gene sequence and decipher whether it is a new species.

The Fungi isolated in the anti Blast gallery belongs to the *Coniophora puteana* species, a frequently described wood-degrader.

4. Conclusion and perspectives

First results show that the LSBB galleries are inhabited by a low diversity of culturable microorganisms with a strong influence of anthropogenic activities. Some results are clearly not predictable by classical thermodynamics, for example the presence of Iron-reducing microorganisms and the presence of methanogenic *Archaea* (for the moment, only detected by their methane production). Further studies including molecular ecology (e.g. high-throughput sequencing of the bulk DNA extracted from the samples, contouring the "Great Plate Count Paradox" [7]) will reveal if the species we isolated are actually representative of the LSBB microflora.

References

[1] E.G. Roussel, M.A. Bonavita, J. Querellou, B.A. Crag, G. Webster, D. Prieur, R.J. Parkes, Sci. **320** (2008).

[2] A.J. Phillips, A.B. Cunningham, R. Gerlach, R. Hiebert, C. Hwang, B.P. Lomans, J. Westrich, C. Mantilla, J. Kirksey, R. Esposito, L. Spangler, Environ. Sci. Technol. **50** (2016).

[3] B.L Maidak, J.R Cole, C.T. Jr. Parker, G.M. Garrity, N. Larsen, N. Li, T.G. Lilburn, M.J. McCaughey, G.J. Olsen, R. Overbeek, S. Pramanik, T.M. Schmidt, J.M. Tiedje, & C.R. Woese, Nucl. Acids Res. **27** (1999).

[4] W.G. Weisburg, S.M. Barns, D.A. Pelletier, & D.J. Lane, J. Bacteriol **173** (1991).

[5] J. Felsenstein, Evol. **39** (1985).

[6] K. Tamura, G. Stecher, D. Peterson, A. Filipski, S. Kumar, Mol Biol Evol. **30** (2013).

[7] R.I. Amann, W. Ludwig, K.H. Schleifer Microbiol Rev. **59** (1995).

Ultra-low Noise EEG at LSBB: New results

Nazanin Hamzei[1], Zoya Bastany[1], Catherine R. Jutzeler[2], Matt Yedlin[1], John L.K. Kramer[3], John D. Steeves[3], and Guy A. Dumont[1,a]

[1] Department of Electrical and Computer Engineering, The University of British Columbia, Vancouver, BC, V6T 1Z4, Canada
[2] Spinal Cord Injury Center, University Hospital Balgrist, University of Zurich, Zurich
[3] International Collaboration On Repair Discoveries (ICORD), Blusson Spinal Cord Centre, 818 West 10th Avenue, Vancouver, BC, V5Z 1M9, Canada

Abstract. In this study, we investigate functional correlates of gamma band oscillations in low-noise EEG signals acquired in the LSBB shielded capsule and compare them to signals acquired in a typical hospital environment. Using a research-grade EEG acquisition system, we acquired 64-channel EEG recordings from three volunteers performing several cognitive, sensory, and motor tasks in both LSBB and hospital settings. Time-frequency analysis on the signals acquired in both environments reveals that the task-induced increase in gamma band (>30 Hz) energy relative to the resting state EEG is more prominent in signals acquired at LSBB, suggesting that task-specific changes in EEG are better reflected and more readily detected in signals acquired at LSBB. These results further demonstrate the potential value of low-noise settings such as the LSBB for conducting challenging high-frequency EEG studies.

1. Introduction

The electroencephalogram (EEG) has proven to be a useful information source in the analysis of brain activity, neural plasticity, and the diagnosis of various neurological disorders, as well as the development of intelligent brain-computer interfaces (BCIs). Advancements in EEG signal acquisition and analysis have enabled improved diagnostic capabilities for seizure detection [1], sleep disorders [2], and Alzheimer's disease [3], just to name a few. EEG has also been extensively used to monitor anesthesia and sedation in the operating room and intensive care unit [4].

EEG can be defined as a representation of neuronal activity (e.g. summated extracellular action potential frequencies) within different brain regions and is usually acquired non-invasively from the surface of the scalp. To reduce the disruptive effects of electrical interference and movement (i.e. noise), depending on the neuronal activity of interest, EEG is often band-pass filtered at different frequencies from 0.5 to 100 Hz and is then studied in terms of activity across different frequency bands; including δ (0–4 Hz), θ (4–7 Hz), α (8–12 Hz), β (12–30 Hz), and γ (30–100 Hz). Clinically, EEG frequencies are most often studied from 0.5 to 30 Hz with the gamma (γ) band being ignored. The gamma band is ignored due to the use of filters to suppress unwanted electromagnetic noise from powered (50 or 60 Hz) electrical outlets. In recent years however, the gamma band has received greater

[a] Corresponding author: guyd@ece.ubc.ca

DOI: 10.1051/e3sconf/20161205003

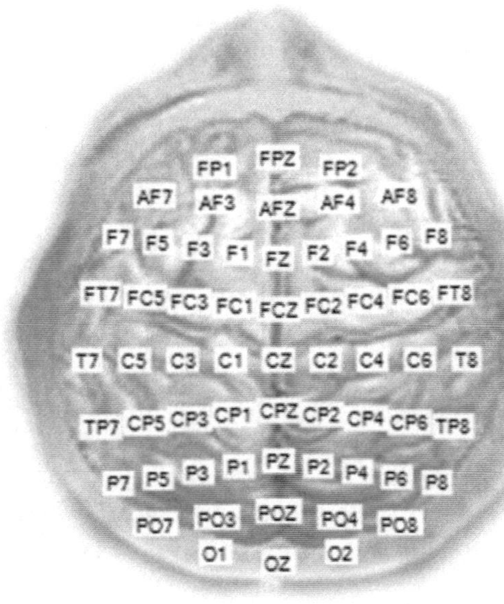

Figure 1. 10–10 electrode montage used for EEG acquisition. Reference and ground electrodes are not shown in the figure.

scrutiny and is typically known to underlie and be modulated by sensorimotor behaviors and internal cognitive processes such as working memory and attention [10].

In this study, we further investigate the potential of the ultra-shielded capsule at the Laboratoire Souterrain Bas Bruit (LSBB in Rustrel, Pays d'Apt, France) for acquisition of clean EEG signals, with a focus on analysis of high frequencies (gamma-band) in search for novel activity patterns. In a previous study, we have shown that EEG signals acquired in LSBB are free from electromagnetic interference and can be acquired and analyzed without the need for notch filtering with battery-powered EEG equipment [6, 7]. This facilitates examination of gamma-band EEG signals acquired in the LSBB while searching for functionally significant high-frequency EEG correlates which would otherwise be obscured by common filters of high-frequency electromagnetic noise. Our objective in this study is to examine the functionally relevant increase in gamma band energy from continuous EEG recordings acquired at LSBB, and test whether gamma-band EEG is accentuated and more readily detected in a low-noise environment than in a typical hospital setting.

2. Data acquisition

The battery-operated NR SIGN EEG 5000Q 64-channel EEG system (NR SIGN Inc., New Westminster, BC, Canada) was used for non-invasive EEG recording (Fig. 1). Data was acquired at a sampling rate of 500 Hz with 16-bit resolution, and was transferred using a USB cable to a laptop computer (also running on battery power) through the NR SIGN EEG application software. The equipment was used to acquire raw, unfiltered continuous EEG recordings at LSBB from seven subjects. Similar recordings were then acquired in ICORD (International Collaboration On Repair Discoveries), a hospital facility within Vancouver General Hospital, using the same equipment and protocols to allow for comparison between the two recording environments. Five subjects had volunteered for data acquisition at ICORD, four of whom were also present during acquisition in LSBB. For the purpose of the study, three subjects were selected out of the four subjects common to both environments, since the acquired data for one subject was found to be corrupted with large movement and muscle artifacts, as well as EOG (high-amplitude electro-oculogram artifact caused by blinking of the eyes) to a high extent.

In each of the environments, the subjects were asked to perform a variety of cognitive, sensory, and motor tasks while EEG was acquired in real time. First, EEG data was collected from the subjects in a darkened room during a relaxed (resting) state as a baseline condition (with either eyes open or eyes closed). The subjects were then asked to perform specific tasks over a 30-second period, while resting for 30 seconds in between the test intervals. The tasks included: 1) counting backwards by 7 from a large randomly selected number; 2) performing an increasingly challenging "matching" memory task on an iPad; 3) performing reciprocal dorsal and plantar flexion movements of the right ankle; and 4) performing repeated flexion-extension of the right wrist. All subjects were right-hand dominant. To assess the effects of pain on the EEG, the subjects were subjected to 5) "brushing" of the adductor pollicis region of the right thumb with a cotton swab as an innocuous tactile stimulation task; and 6) application of a hot pack (recorded skin temperature = 34-36°C) to the same adductor pollicis region as a noxious stimulus, while rating their pain perception using the (0–10, 10 being most painful) visual analog scale (VAS). For the latter task, only two out of the three subjects were included in the analysis since one subject did not continue the recording at the hospital environment, claiming the stimulus was too hot to tolerate.

3. Data analysis

3.1 Pre-processing

All recordings were visually inspected in the time domain for removal of periods containing broadly distributed, high-amplitude movement artifacts, electrode saturation, and other irregular artifacts. Electrodes with poor skin contact, judged by their abnormal activity patterns throughout the length of the recording, were removed from the data at each of the recordings and the analysis was based on stable channel recordings. Occurrence of eye blinks and eye movements were not criteria for data rejection since their frequency content is usually below 15 Hz and did not overlap with the gamma band frequencies. Naturally, there was no need for application of notch filters to recordings acquired in LSBB. As for recordings at ICORD, no notch filer was used either owing to the fact that the battery-operated acquisition system was also able to minimize external electromagnetic power-line interference automatically using well-calibrated differential amplifiers.

3.2 Time-frequency analysis

In order to quantitatively measure the oscillatory activity of EEG signals, time-frequency spectrograms were computed using the Stockwell transform [8]. Also known as the S-transform, the Stockwell transform is a variant of the continuous wavelet transform (CWT) which uses a moving and scalable localizing Gaussian window. While having the advantage of frequency-dependent resolution over Short-time Fourier Transform (STFT), the S-transform is also shown to have a frequency invariant amplitude response, contrary to the Continuous Wavelet Transform which forces more attenuation on high frequencies. These features make the S-transform an ideal tool for detection of high-frequency activity in EEG signals. The S-transform of a signal $x(t)$ is given by:

$$S(\tau, f) = \int_{-\infty}^{+\infty} x(t)|f|\frac{e^{\frac{(t-\tau)^2 f^2}{2}}}{\sqrt{2\pi}} e^{-2\pi i f t} dt \qquad (1)$$

 DOI: 10.1051/e3sconf/20161205003

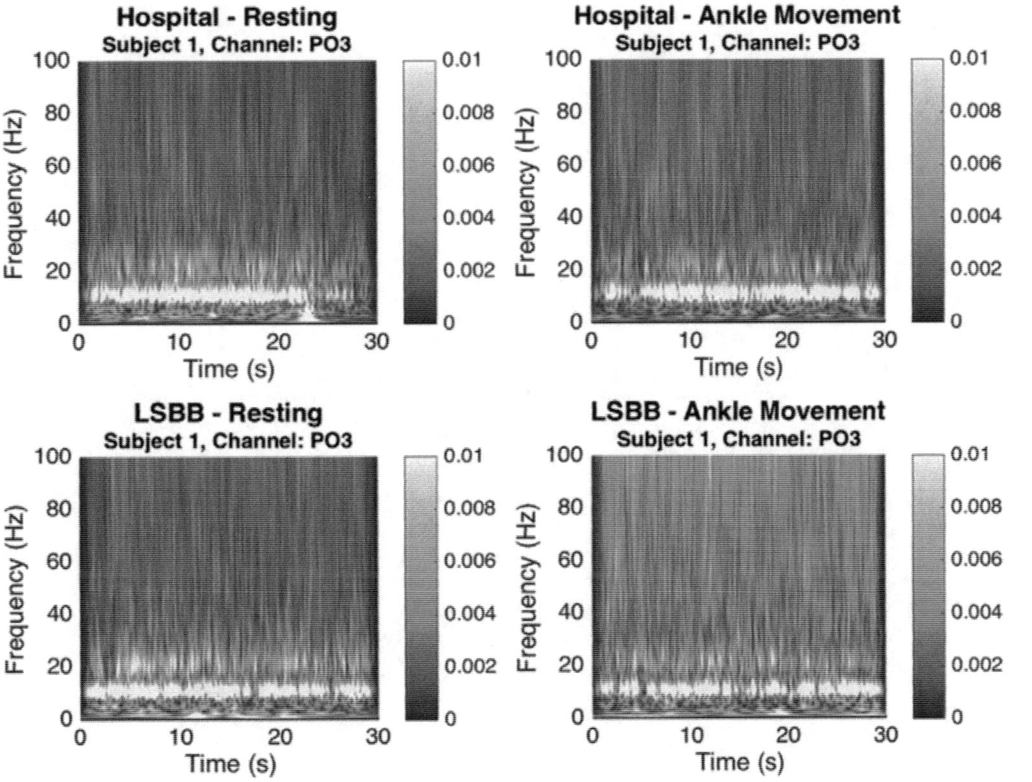

Figure 2. Magnitude of the Stockwell transform of normalized EEG signals from one channel (PO3) of a subject during rest (left column) and ankle movement (right column), acquired at the hospital facility (top row) versus the LSBB (bottom row). Task-specific changes within gamma band are more significant and easier to detect in LSBB.

Figure 2 shows the magnitude of the S-transformed EEG signals of one subject during 30-second intervals of rest and a repetitive motor activity (ankle movement) in the different recording environments. Comparing the figures on the left (resting state) with those on the right (motor task), we observe that the motor activity has caused power increases within the gamma band (f >30 Hz); however, this increase in high-frequency power is more apparent in the LSBB when compared to the hospital environment.

To quantitatively assess the task-specific changes in the gamma band EEG, we calculated the task-rest gamma band energy ratios of the S transform and compared them between LSBB and the hospital environment. To this end, artifact-free recordings for all subjects and all tasks were first segmented into consecutive epochs of 30 seconds. Each epoch was then normalized in order to reduce the influence of electrode impedance and alleviate the disparities across subjects due to factors such as skin conductivity, which yielded epochs with comparable amplitude values. Channel-wise normalization was performed in the time domain by removing the mean and dividing by the root-mean-square value of the corresponding channel [4], according to the equation below:

$$x_N(t) = \frac{x(t) - \bar{x}}{x_{rms}} \tag{2}$$

Table 1. Grouping of electrodes based on different sites of the brain.

Brain Region	Electrodes
Prefrontal	FPZ, FP1, FP2, AFZ, AF3, AF4, AF7, AF8
Frontal	FZ, F1, F2, F3, F4, F5, F6, F7, F8, FCZ, FC1, FC2, FC3, FC4
Central	CZ, C1, C2, C3, C4, CPZ, CP1, CP2, CP3, CP4
Left Temporal	FC5, FT7, C5, T7, CP5, TP7
Right Temporal	FC6, FT8, C6, T8, CP6, TP8
Left Parietal	P3, P5, P7, PO3, PO7
Right Parietal	P4, P6, P8, PO4, PO8
Parietal	PZ, P1, P2, POZ
Occipital	OZ, O1, O2

We measure the high frequency activity during each of the epochs by computing the total gamma band energy of the epoch in the time-frequency space, defined as:

$$E_{\gamma} = \int_{f_L=30}^{f_H=100} \int_{\tau_i=0}^{\tau_f=30} |S_N(\tau, f)|^2 d\tau df. \tag{3}$$

Where $|S_N(\tau, f)|$ refers to the magnitude of the S transform of the normalized signal (Eq. (2)), and integration is performed over the entire epoch length (30 seconds) and within the gamma band (30–100 Hz). Consequently, responses to each task for each subject were expressed as task-rest ratios; that is, given there are N epochs corresponding to the rest condition and M epochs belonging to a particular task, we compute a total of $M * N$ ratios by dividing the total gamma band energy of each task epoch by the total gamma band energy of each of the resting-state epochs. We then compare the entire set of ratios (including all subjects) at nine functionally different brain regions (see Table 1) in LSBB with that of the hospital environment.

4. Discussion

The results from time-frequency analysis across all subjects in both environments are summarized in Fig. 3, where each sub-plot corresponds to a specific grouping of electrodes (brain regions) according to Table 1. We have grouped the tasks into three categories: cognitive tasks (Fig. 3a) including counting and matching, motor tasks (Fig. 3b) including right ankle movement and right wrist movement, and right hand sensory tasks (Fig. 3c) including brushing and application of a noxious heat pack to the dorsal surface of the right thumb. Tasks belonging to each category are shown alongside each other. For each task, we have shown a boxplot of the task-rest ratios acquired in the hospital environment (left) along with a boxplot of the ratios acquired at LSBB. Shown in the plots are the median value, as well as the 25% and 75% quartiles, and the whiskers representing $\pm 2.7\sigma$ or 99.3 percent coverage given the data is normally distributed. These plots allow for comparison of the distribution of gamma band task-rest ratios in LSBB and hospital across different brain regions.

A quick look at Fig. 3b reveals that for both of the motor tasks, the gamma band motor-rest ratios are significantly larger and more readily detected in LSBB across all regions of the brain. While the median of motor-rest ratios at LSBB is always greater than one, the median of the ratios at the hospital stays close to one, suggesting that the motor tasks were not readily detectable at the hospital. This is clearly an advantage for studies in a low electromagnetic noise environment like the LSBB, indicating its potential for detection of EEG benchmarks related to understanding central nervous system (CNS) control of basic motor tasks or early detection of disorders associated with changes in motor behaviors (e.g. Epilepsy, Parkinson's disease, etc.). Low-noise environments, like the LSBB, would also

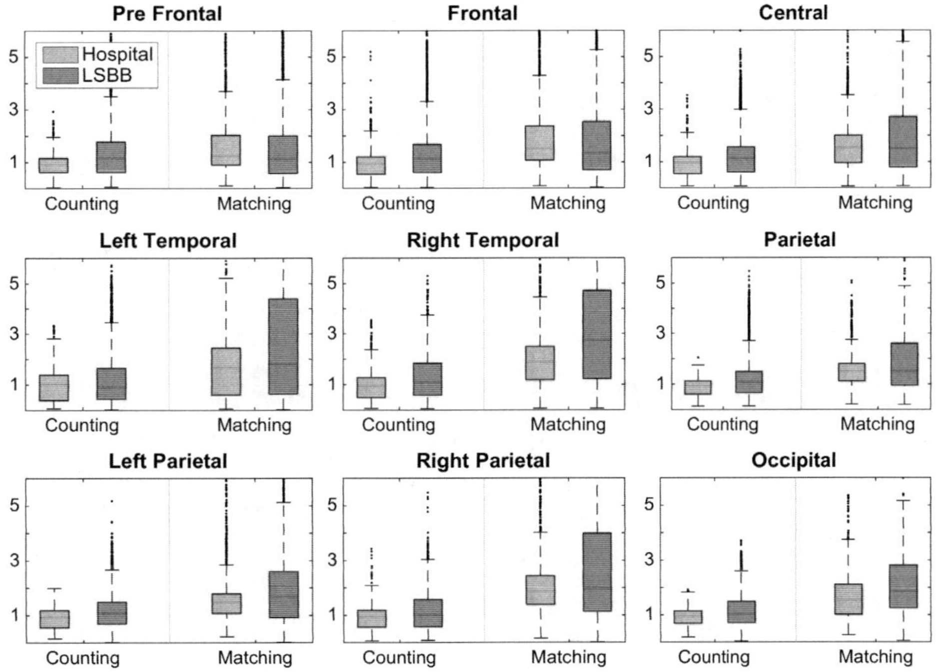

(a)

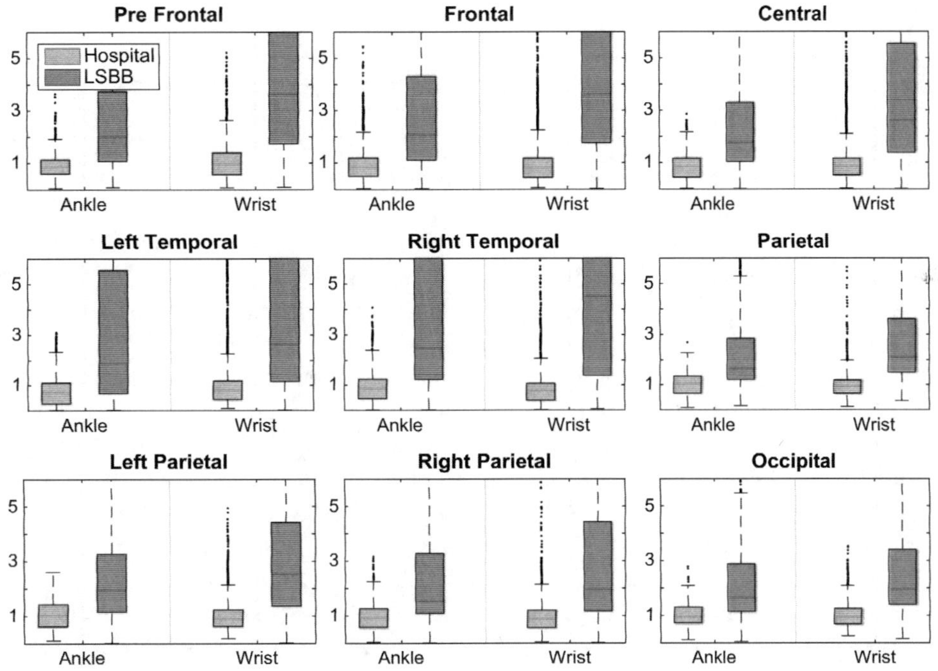

(b)

Figure 3. Comparison of distribution of task-rest ratios between signals acquired at LSBB and the hospital at different brain regions based on the grouping scheme outlined in Table 1. a) Cognitive tasks including counting and matching, b) motor tasks including ankle and wrist movements, c) sensory tasks including brushing and application of heat pack.

 DOI: 10.1051/e3sconf/20161205003

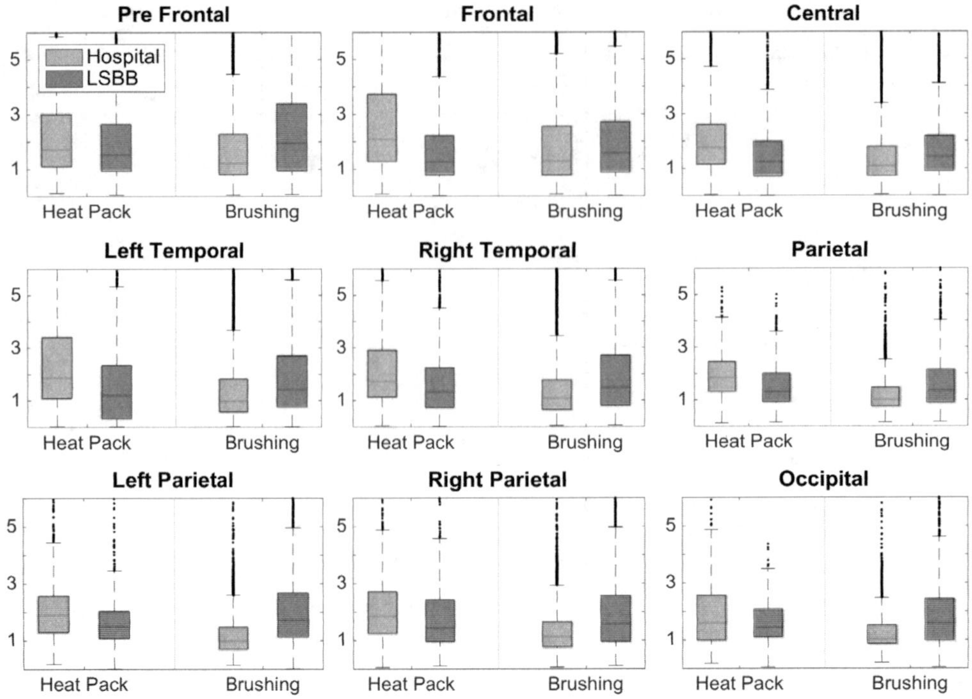

(c)

Figure 3. Continued.

facilitate a greater understanding for potential subtle changes in cortical plasticity after CNS motor injuries (e.g. spinal cord or brain damage).

We have depicted the ratios for cognitive tasks in Fig. 3a. Although having slightly higher values in LSBB, backward counting-rest ratios are mainly distributed around one in both environments, suggesting that gamma band may not be significantly involved in continuous counting tasks. We have found this to be the case also in a previous study [6]. On the other hand, matching-rest ratios are mainly greater than one in both environments, and their values are higher in LSBB at temporal and occipital regions of the brain. As an explanation for why a global increase in ratios is not seen at LSBB, it could be claimed that the increase in gamma band due to matching is more focused, contrary to the motor functions which exhibit a more global cortical response. Thus, in brain regions where the gamma-band content is boosted due to the cognitive activity associated with the matching task, the ratio increase is more prominent within the low-noise environment of the LSBB than the hospital environment. This hypothesis is backed by the fact that temporal lobes are known to be associated with visual memory and object recognition [9], two functions that are specifically needed for a matching task.

Brushing-rest and noxious heat-rest ratios were also compared between the two environments (Fig. 3c). This figure demonstrates that the median of brushing-rest ratios is slightly higher in LSBB compared to the hospital, thus revealing the potential of LSBB in identifying the brushing-induced high frequency activity. However, the heat-rest ratios are lower in LSBB than in the hospital environment. Explaining this issue, it is worth mentioning that the average of the VAS pain ratings across all recordings and all subjects was 5.5 at the hospital and 1.5 at LSBB, suggesting that despite the efforts to keep the experiment conditions equal, the subjects experienced more pain in the hospital facility than in LSBB. We might therefore hypothesize that the lower ratios in LSBB are caused by the lower pain experience in comparison with ICORD, which would in turn yield lower gamma band activity in LSBB as the strength of gamma band oscillations has been shown to correlate well with the intensity of the perceived pain [10]. However, since only a small number of subjects (two) were included for the heat task, caution must be exercised in interpreting the results pertaining to this task.

5. Conclusion

We have summarized a small pilot study on the potential of an ambient low-noise environment (e.g. LSBB) for the easy identification of task-specific high-frequency gamma-band EEG signals. Our results demonstrate that functionally correlated gamma-band EEG patterns can be better detected in low-noise conditions when compared with a typical hospital environment. This motivates the design of more informative studies with the end goal of defining potentially novel and predictive high-frequency EEG (gamma-band benchmarks) for a better understanding of central neuronal function and CNS disorders.

To address the limitations of the current work, studies using a larger number of subjects are necessary for more stringent statistical validation and more confident conclusions. One might also consider designing a protocol for detection of event-related potentials (ERPs) with time-locked experimental trials as opposed to spontaneous EEG, in order to more thoroughly investigate the role of gamma-band EEG activity in task-specific sensorimotor behaviors and/or cognitive activities.

References

[1] T.N. Alotaiby, S.A. Alshebeili, T. Alshawi, I. Ahmad, F.E. Abd El-Samie, *EEG seizure detection and prediction algorithms: a survey*, EURASIP Journal on Advances in Signal Processing **2014**, 1–21 (2014).

[2] S. Kubicki, W. Scheuler, H. Wittenbecher, *Short-term sleep EEG recordings after partial sleep deprivation as a routine procedure in order to uncover epileptic phenomena: an evaluation of 719 EEG recordings*, Epilepsy Res. Suppl., **2**, 217–230 (1991).

[3] J. Jeong, *EEG dynamics in patients with Alzheimer's disease*, Clinical neurophysiology, **115**, 1490–1505 (2004).

[4] T. Zikov, S. Bibian, G.A. Dumont, M. Huzmezan, C.R. Ries, *Quantifying cortical activity during general anesthesia using wavelet analysis*, IEEE Transactions on Biomedical Engineering, **53**, 617–632 (2006).

[5] M.W. Howard, D.S. Rizzuto, J.B. Caplan, J.R. Madsen, J. Lisman, R. Aschenbrenner-Scheibe, A. Schulze-Bonhage, M.J. Kahana, *Gamma oscillations correlate with working memory load in humans,* Cerebral cortex **13**, 1369–1374 (2003).

[6] A.S. Zandi, G.A. Dumont, M.J. Yedlin, P. Lapeyrie, C. Sudre, S. Gaffet, *Scalp EEG Acquisition in a Low-Noise Environment: A Quantitative Assessment*, IEEE Transactions on Biomedical Engineering, **58**, 2407–2417 (2011).

[7] A.S. Zandi, G.A. Dumont, M.J. Yedlin, P. Lapeyrie, C. Sudre, S. Gaffet, *Analysis of scalp EEG recorded in a low-noise environment*, EDP Sciences (2011).

[8] R.G. Stockwell, L. Mansinha, R.P. Lowe, *Localization of the complex spectrum: the S transform*, IEEE Transactions on Signal Processing, **44**, 998–1001 (1996).

 DOI: 10.1051/e3sconf/20161205003

[9] N. Axmacher, D.P. Schmitz, T. Wagner, C.E. Elger, J. Fell, *Interactions between medial temporal lobe, prefrontal cortex, and inferior temporal regions during visual working memory: a combined intracranial EEG and functional magnetic resonance imaging study*, The Journal of Neuroscience **28**, 7304–7312 (2008).

[10] Z.G. Zhang, L. Hu, Y.S. Hung, A. Mouraux, G.D. Iannetti, *Gamma-band oscillations in the primary somatosensory cortex—a direct and obligatory correlate of subjective pain intensity*, The Journal of Neuroscience **32**, 7429–7438 (2012).

Noise modeling in a signal conditioning circuit for low power audio application using resistive sensor

Stéphane Meillère[1], Eric Savary[1], Wenceslas Rahajandraibe[1], and Hervé Barthélémy[2]

[1] IM2NP UMR CNRS 7334, Aix Marseille University, France
[2] IM2NP UMR CNRS 7334, University of Toulon, France

Abstract. Piezoresistive sensors convert a physical value into a resistance variation. Often four resistive elements are connected together in a Wheatstone bridge to provide electrical variations of sensors. When this structure is biased with a fixed voltage source or a current source the topology provides a differential output voltage. To exploit information a conditioning circuit is associated to the bridge. In most cases it consists of an instrumentation amplifier followed by a data converter to obtain very quickly a digital representation of information. Due to the high input impedance of the instrumentation amplifier, bridge sensitivity is preserved. A filter may be added to avoid aliasing or a continuous time sigma-delta modulator that includes filtering feature. This study is concerning the conditioning structure for piezoresistive sensors bridge especially fully integrated microphones for biomedical application. The bridge signal to noise ratio is set by biasing the amplifier stage by current. The noise performance becomes the limiting factor of the read-out circuit. Current mode topologies drive amplifiers design where inputs are the main noise contributor. Modeling noise contribution is a key point in the design of the conditioning circuit. The current consumption leads noise performances too. A proposed architecture was implemented in a 65nm CMOS standard technology for performance measurement and evaluation with nanowire based microphone dedicated to hearing aids application.

1. Introduction

The application of a cochlear implant needs to develop amplifiers for conditioning circuits of nano scale M&NEM (Micro & Nano Electro Mechanical) sensors. Those sensors are microphones. The amplifier of the sensor necessarily adds noise to the useful signal. Thereby, the resulting signal to noise ratio SNR is necessarily less than a single sensor. The evaluation of performance in terms of EIN for the entire microphone used to put these performances in the prior art regardless of the nature of the output signal (digital or analog) by comparing sound pressure levels expressed in dB SPL (Signal Pressure Level) common unit at any microphones. These considerations lead to the development of the curve shown in the graph of Fig. 1 for a bias current of 100 μA acoustic noise reduced in set entry is indicated as a function of the noise added by the amplifier. This curve can anticipate the performance of the system in terms of acoustic noise at the input according to the noise added by the amplifier.

The equivalent acoustic noise levels of MEMS microphones already marketed are indicated along the ordinate axis [1, 2]. The performance of CMOS amplifiers described

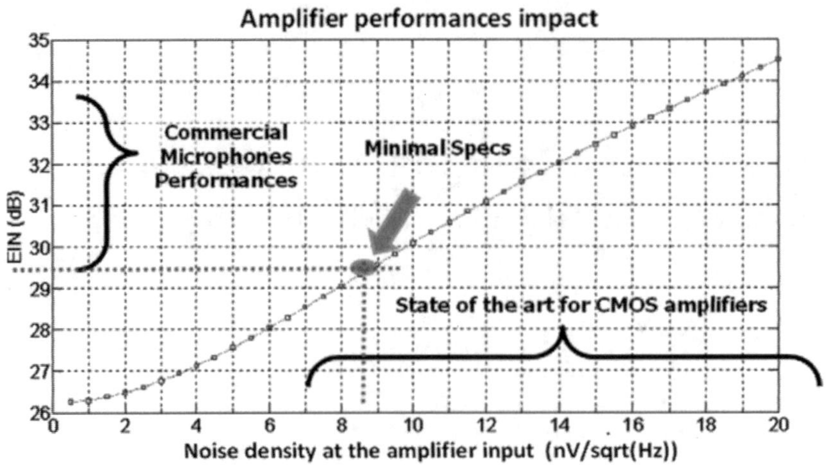

Figure 1. Noise density at the amplifier input versus EIN.

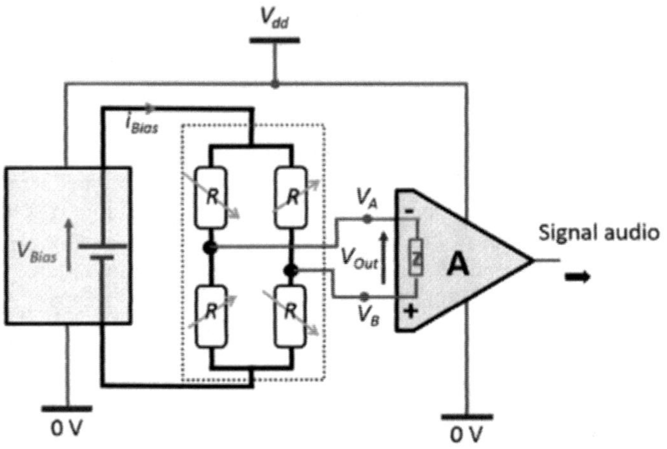

Figure 2. Conditioning circuit.

in the prior art are indicated along the abscissa axis. This shows that it is possible to supplant MEMS microphones with existing amplification circuits [3–5].

2. Conditioning circuit

The detailed study of the operation of the sensor has to highlight the specific constraints bridge itself as the presence of common mode voltage and output impedance. The simplified topology of the conditioning circuit of the sensor is shown in Fig. 2:

Biasing involves power consumption. In the simplest case, if the constraints are not so strong the sensor can be directly polarized with the same power as the amplifier. This implies that the equivalent resistance R_0 for the association of four gauges is high enough to maintain reasonable power loss, or that the supply voltage is sufficiently small [5–9]. However, current integrated circuits described in the state of the art as well as those sold in the industry in general operate with supply voltages above 1.2V. This supply applied to the sensor described above gives rise to a current I_{Bias} equal to 279uA. This corresponds to a power of 334mW. This power is excessive due to what it can not be dissipated by the nanoscale sensors and it

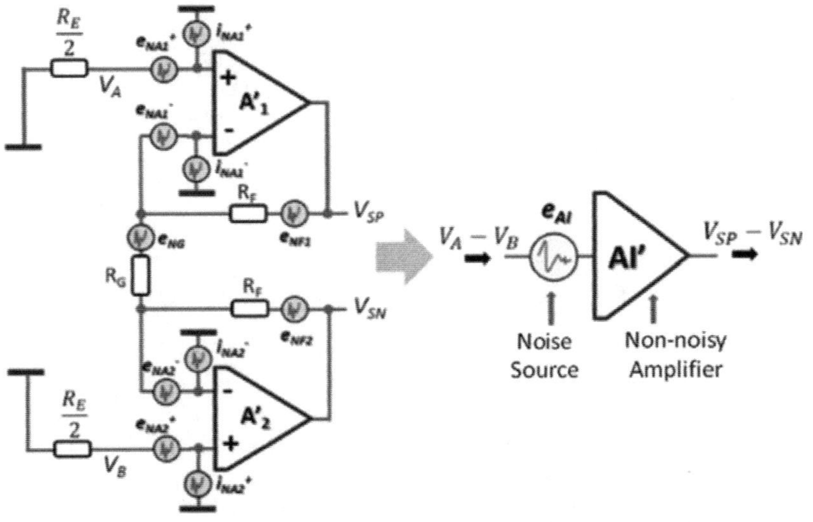

Figure 3. Noise sources contribution in the instrument amplifier.

causes an incompatible use with biomedical targeted application. It is therefore essential to have a mastering circuit for the polarization regardless of the supply voltage of the amplifier. The common mode output voltage of the sensor can cause problems to the amplifier when the assembly is referenced to 0V. Indeed, for an I_{Bias} current of 100 μA, the bias voltage V_{Bias} is 0.4V. In this case, the common mode voltage V_{CM} is 0.2V. This means the input signals V_A and V_B of the amplifier operate around a mean value of 0.2V. In order to drive this common mode voltage to a value compatible with the amplifier it is essential to drive independently the current V_{Bias} from the common mode voltage. The difficulty at that time is to determine the correct noise influence of all elements in this system and especially the noise contribution of the instrument amplifier used.

3. Noise modeling for the instrument amplifier

The overall contribution of the instrumentation amplifier shown is mainly determined by the input stage. The diagram in Fig. 3 shows the noise model of the first order. This model is composed by non-noise amplifiers rated A'_1 and A'_2. They are completed by voltages sources at inputs $e_{NA1}{}^+$, $e_{NA1}{}^-$ and $e_{NA2}{}^+$, $e_{NA2}{}^-$ and the current sources $i_{NA1}{}^+$, $i_{NA1}{}^-$ and $i_{NA2}{}^+$ and $i_{NA2}{}^-$. All these sources represent the noise contribution in real circuits. The contribution of resistance is modeled by the voltage sources e_{NF} (1 and 2) and e_{NG}. Assuming that noise sources are not correlated and that the amplifiers A'_1 and A'_2 are identical, it is possible to represent all the contributions in the form of a single noise source e_{AI} so it reduced the input of an amplifier AI' not noisy. So, the equivalent noise model will compare the impact of different noise sources in the amplification structure. We are not able to sum directly the instantaneous voltages and the instantaneous currents of noise [11–13] due to their specific natures (random fluctuation), the influence of each contribution will be determined separately.

The power of the source of the noise source e_{AI} at the input will be determined by considering the sum of the noise power output of the amplifier stage. The noise model is decomposed as shown in Figs. 4a–d.

The voltages V_{IPA1}, V_{INA1}, V_{EPA1}, V_{ENA1} sources show the noise contributions at the input in amplifier A'_1. The noise figure expressions $i_{NA1}{}^+$, $i_{NA1}{}^-$, $e_{NA1}{}^+$ and $e_{NA1}{}^-$ are shown in

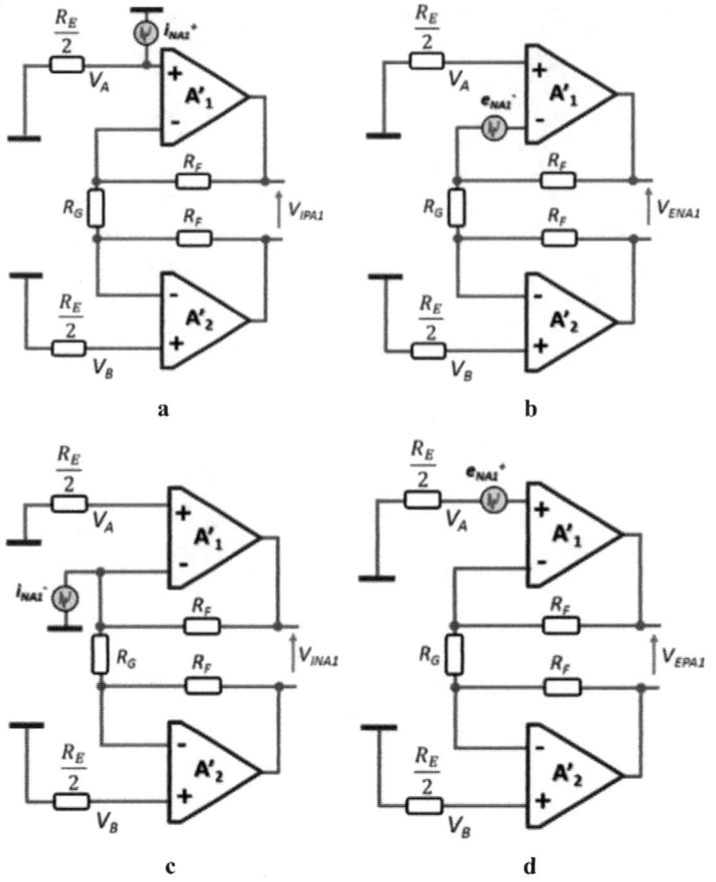

Figure 4.

Figs. 4a–d and are expressed in Eqs. (1)–(4) respectively.

$$V_{IPA1} = i^{+}_{NA1}\frac{R_E}{2}\left(1+\frac{2R_F}{R_G}\right) \tag{1}$$

$$V_{INA1} = -i^{-}_{NA1}R_F \tag{2}$$

$$V_{EPA1} = \frac{e^{+}_{NA1}}{2}\left(1+\frac{2R_F}{R_G}\right) \tag{3}$$

$$V_{ENA1} = \frac{e^{-}_{NA1}}{2}\left(1+\frac{2R_F}{R_G}\right). \tag{4}$$

The voltages V_{IPA2}, V_{INA2}, V_{EPA2}, V_{ENA2} sources show the noise contributions at the input in amplifier A'_2. The expressions of the noise sources are obtained by following the same way and it leads to Eqs. (5)–(8).

$$V_{IPA2} = i^{+}_{NA2}\frac{R_E}{2}\left(1+\frac{2R_F}{R_G}\right) \tag{5}$$

$$V_{INA2} = -i^{-}_{NA2}R_F \tag{6}$$

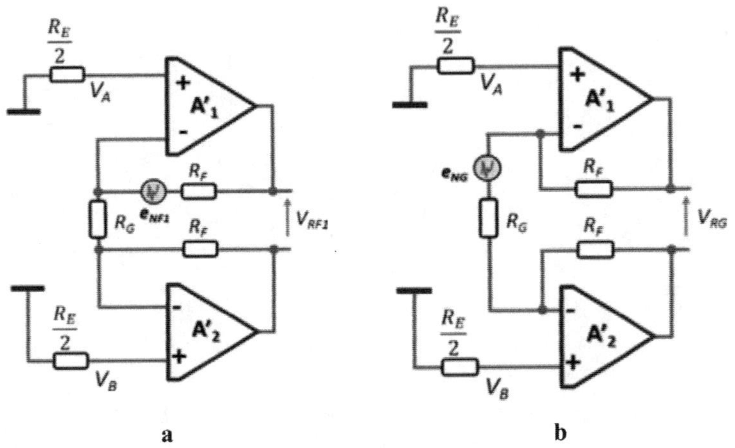

Figure 5.

$$V_{EPA2} = \frac{e^{+}_{NA2}}{2}\left(1+\frac{2R_F}{R_G}\right) \tag{7}$$

$$V_{ENA2} = \frac{e^{-}_{NA2}}{2}\left(1+\frac{2R_F}{R_G}\right). \tag{8}$$

The resistors also make a noise contribution as shown in Figs. 5a and b. The voltages V_{RF1}, V_{RF2} and V_{RG} illustrate the voltage sources contribution of the noisy resistances in the amplifier.

$$V_{RF1} = e_{NF1} \tag{9}$$

$$V_{RF2} = e_{NF2} \tag{10}$$

$$V_{RG} = -e_{NG}\left(\frac{2R_F}{R_G}\right). \tag{11}$$

After the establishment of the noise sources, it is possible to calculate the total noise power V^2_{NOISE} as the sum of the powers of each component. When the amplifiers A'_1 and A'_2 are identical the expressions (12) to (16) give the noise power of each contribution.

$$i^{+\,2}_{PA1} = i^{+\,2}_{PA2} = i_{PA}{}^2 \rightarrow V_{IPA1}{}^2 = V_{IPA2}{}^2 = V_{IPA}{}^2 \tag{12}$$

$$i^{+\,2}_{NA1} = i^{+\,2}_{NA2} = i_{NA}{}^2 \rightarrow V_{INA1}{}^2 = V_{INA2}{}^2 = V_{INA}{}^2 \tag{13}$$

$$e^{+\,2}_{NA1} = e^{+\,2}_{PA2} = e_{NA}{}^2 \rightarrow V_{EPA1}{}^2 = V_{EPA2}{}^2 = V_{EPA}{}^2 \tag{14}$$

$$e^{-\,2}_{NA1} = e^{-\,2}_{PA2} = e_{NA}{}^2 \rightarrow V_{ENA1}{}^2 = V_{ENA2}{}^2 = V_{ENA}{}^2 \tag{15}$$

$$e_{NF1}{}^2 = e_{NF2}{}^2 = e_{NF}{}^2 \rightarrow V_{RF1}{}^2 = V_{RF2}{}^2 = V_{RF}{}^2. \tag{16}$$

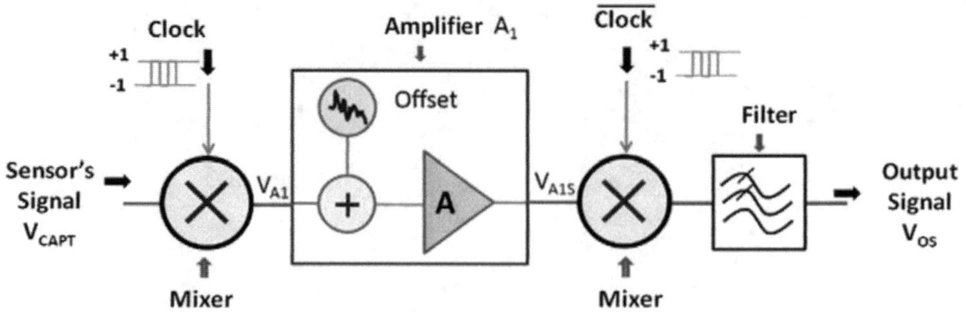

Figure 6. Chopper noise rejection technique.

The noise power V^2_{NOISE} output of the amplifier is expressed in (17).

$$V_{NOISE}{}^2 = 2V_{IPA}{}^2 + 2V_{INA}{}^2 + 2V_{EPA}{}^2 + 2V_{ENA}{}^2 + 2V_{RF2}{}^2 + V_{RG}{}^2. \quad (17)$$

This Eq. (18) determines the equivalent noise power e^2_{AI} of this amplification stage.

$$e_{IA}{}^2 = 2i_{NA}{}^2\left[\left(\frac{R_E}{2}\right)^2 + \frac{R_F{}^2 R_G}{R_G + 2R_F}\right] + 4e_{NA}{}^2$$

$$+2e_{NF}{}^2\frac{R_G}{R_G + 2R_F} + e_{NG}{}^2\left(\frac{2R_F}{R_G + 2R_F}\right)^2. \quad (18)$$

When the voltage gain is important, ie $R_F \gg R_G$, this expression can be approximated and gives in (19).

$$e_{IA}{}^2 = 2i_{NA}{}^2\left[\left(\frac{R_E}{2}\right)^2 + \frac{R_F R_G}{2}\right] + 4e_{NA}{}^2 + e_{NG}{}^2. \quad (19)$$

The noise voltage added by the amplifier is the most significant regarding the other element. Indeed, the resistor R_F is a negligible contribution and the resistance R_G can be chosen as small as possible for example 100Ω and this gives noise density of 0.407nV/sqrt(Hz). This rough estimate does not involve the frequency characteristic of the amplifier or the frequency distribution of the noise. However, it can be shown that the noise current contribution i_{NA} plays an important role through the impedance of the sensor. It will be necessary to pay attention to the noise current generated by the amplifier and the output impedance of the sensor R_E.

4. Conditioning circuit improvement

The structure presented on Fig. 6 the technique called chopper [12]. This is commonly used because it eliminates the offset of an amplifier and these aberrations that are somehow comparable to a very noise low frequency.

At the beginning of system, the V_{CAPT} signal from the sensor is modulated by an alternative clock signal of 1 or −1 in a first mixer. Therefore, the input of the amplifier A_1, sees the VCAPT sensor signal during the period of the mixer clock. So, it results the Eqs. (20) and (21).

$$V_{A1} = 1 \cdot V_{CAPT} \quad (20)$$

$$V_{A1} = -1 \cdot V_{CAPT}. \quad (21)$$

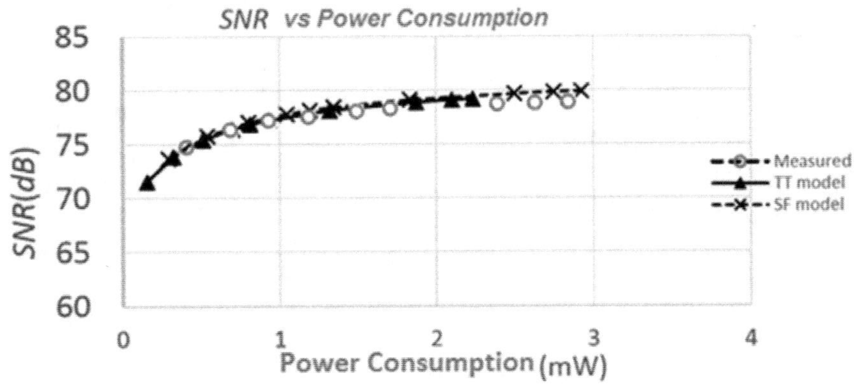

Figure 7. SNR of the amplifier versus power consumption.

The output of the amplifier, the signal V_{A1S} and offset added by the amplifier are magnified by a factor k_v as shown in the expressions (22) and (23):

$$V_{A1S} = k_v \cdot \left(V_{CAPT} + V_{off}\right) \tag{22}$$

$$V_{A1S} = k_v \cdot \left(-V_{CAPT} + V_{off}\right). \tag{23}$$

The output of the amplifier, the signal is demodulated by a second mixer. The result is a voltage composed of the useful signal affected by a positive offset and a negative offset.

$$V_{A1M} = k_v \cdot V_{CAPT} + k_v \cdot V_{off} \tag{24}$$

$$V_{A1M} = k_v \cdot V_{CAPT} - k_v \cdot V_{off}. \tag{25}$$

In the end of this system, a low-pass filter completes the process and permits an offset reduction by providing the assessment of the average value of the signal. Assuming that the clock is perfectly symmetrical this average is given in (26).

$$\overline{V_{OS}} = k_v \cdot V_{CAPT}. \tag{26}$$

This technique is also applied to reduce the impact of noise at low frequency in CMOS technology [13]. The modulation can be viewed as a way to move the spectral content of the useful signal to a frequency band in which the amplifier generates less noise. After amplification, the second frequency conversion, replace the useful signal to its original band and shifts the low frequency amplifier noise outside the useful band. This one is removed at the end of string by the low-pass filter.

5. Results and analyses

The Fig. 7 shows the signal to noise ratio versus power consumption and Fig. 8 illustrates the DC bias current in the amplifier versus de biasing voltage. These results are given for measured and simulated with CADENCE Virtuoso simulator with a 0.65 μm CMOS standard process. It validates the noise modeling of our amplifier. The TT, SF, FF, SS models are typical, slow-fast, fast-fast and slow-slow corners variations of the CMOS process.

The noise and bias current are well determinate and validated by measurements. The simulations suit to measurements. In contrast, the noise added by the amplifier is more significant. To determine the most favorable topology, the impact of noise contributors in the voltage topology was studied. The Fig. 9 presents a die photo of an all integrated conditioning

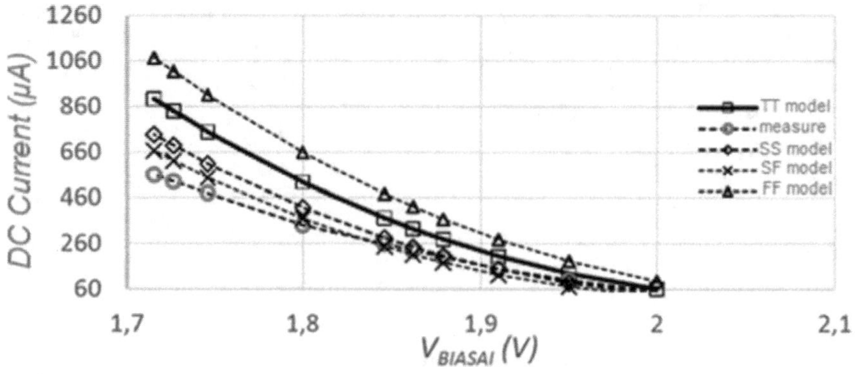

Figure 8. DC bias current of the amplifier versus BIAS voltage.

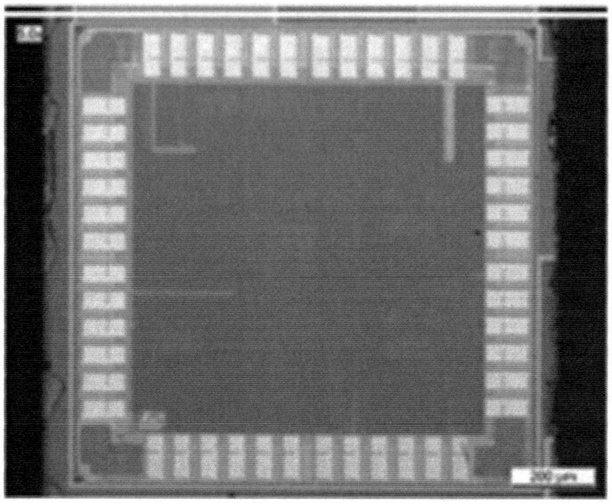

Figure 9. Die area of approximate 1 mm^2.

circuit for MEMS sensors. This circuit was realized in a 0.65 μm standard CMOS and the die area is approximately 1mm^2.

6. Conclusions

In this paper we have demonstrated a model of noise inside a conditioning circuit dedicated to nano scale sensors. It appears in this study an optimum bias current to achieve the desired performance and a significant constraint on the noise added by the conditioning circuit. Indeed, it appeared to the court of the study from the constraints of conditioning circuit as the noise added by circuits around the sensor is a major constraint. Only too sensitive to common mode topologies have been excluded due to the excess of sensibility. In order to make a choice in relation to the requirements of the application a brief study of the impact of noise contributors present in amplifiers was conducted for the voltage topologies and power topologies. Some additional measurements need to be involved with the LSBB partnership. The campaign might determinate the correct noise floor of our system and great measurement results without any external and additional noises.

References

[1] Microphones, M., Technologies for High Performance Portable Healthcare Devices.

[2] Verdot, T., et al., Modélisation d'une architecture innovante de microphone MEMS à détection par nano jauges. 21ème Congrès Français de Mécanique, 26 au 30 août 2013, Bordeaux, France (FR), 2013.

[3] Czarny, J., et al. New architecture of MEMS microphone for enhanced performances. in Semiconductor Conference Dresden-Grenoble (ISCDG), 2013 International, 2013, IEEE.

[4] Marton, L. and C. Marton, ADVANCES ELECTRONC & ELECTRON PHYSICS, 1980, Elsevier Science.

[5] Ghallab, Y.H. and W. Badawy, A new topology for a current-mode wheatstone bridge. Circuits and Systems II: Express Briefs, IEEE Transactions on, 2006, **53**(1): p. 18–22.

[6] O'Grady, A., Transducer/sensor excitation and measurement techniques. Analog Dialogue, 2000, **34**(5): p. 1–6.

[7] Peyton, A. and V. Walsh, Analog Electronics with Op-amps: A Source Book of Practical Circuits, 1993, Cambridge University Press.

[8] Azhari, S.J. and H. Kaabi, AZKA cell, the current-mode alternative of Wheatstone bridge. Circuits and Systems I: Fundamental Theory and Applications, IEEE Transactions on, 2000, **47**(9): p. 1277–1284.

[9] Toumazou, C. and F. Lidgey, Novel current-mode instrumentation amplifier. Electronics Letters, 1989, **25**(3): p. 228–230.

[10] Barthelemy, H., et al. CMOS instrumentation-amplifier based on ASKA cell. in NEWCAS Conference (NEWCAS), 2010 8th IEEE International, 2010.

[11] Jaquay, J.W., Designers Guide to: Instrumentation amplifiers. Experimental Techniques, 1977, **2**(2): p. 40–43.

[12] Enz, C.C. and G.C. Temes, Circuit techniques for reducing the effects of op-amp imperfections: autozeroing, correlated double sampling, and chopper stabilization. Proceedings of the IEEE, 1996, **84**(11): p. 1584–1614.

[13] Jimmin, C., A.A. Abidi, and C.R. Viswanathan, Flicker noise in CMOS transistors from subthreshold to strong inversion at various temperatures. Electron Devices, IEEE Transactions on, 1994, **41**(11): p. 1965–1971.

On the investigation of voltage controlled oscillator phase noise for IoT applications

F. Haddad, W. Rahajandraibe, and I. Ghorbel

Aix-Marseille University, IM2NP UMR CNRS 7334, Technopole château Gombert, 13013 Marseille, France

Abstract. Voltage controlled oscillator (VCO) is one of the key elements in radio frequency (RF) transceivers. A VCO working at 2.4 GHz and designed in CMOS technology is presented. It is suitable for low-cost and low-noise applications using wireless standards such as ZigBee, Bluetooth, Wi-Fi and WPAN (Wireless Personal Area Network). The noise characteristics of this RF VCO are investigated. Noise measurements, especially, phase noise are achieved under different environmental conditions.

1. Introduction

The last few years have seen a dramatic rise in Internet of Things (IoT) devices and connected products such as wireless sensors, home automation systems, smart meters, wearable, etc. [1]. Gartner estimates that 4 billion connected things will be in use in the consumer sector in 2016, and will reach 13.5 billion in 2020 [2]. The network topology of IoT systems consists of nodes that collect and transmit a limited amount of data to a central controller. Figure 1 presents the architecture of a typical IoT sensor node. It is comprised of a processor unit or a microcontroller (MCU) that processes data and runs software stacks interfaced to a wireless device for connectivity.

Sensor nodes should be portable and able to work under batteries or even able to harvest ambient energy for several years. The main challenge consists in minimizing power consumption and extending wireless connectivity range. To this end, several low-power and low-cost narrow-band standards, such as IEEE 802.15.4 (LR-WPAN) [3] and Bluetooth Low Energy [4] have emerged.

The wireless transceiver is an important part of IoT systems. It achieves transmission (TX) and reception (RX) of the data using radio frequency (RF) connectivity (cf. Fig. 1). The 2.4 GHz frequency is a good choice to serve the transceiver designer needs, given the high profile of this 2.4 GHz ISM (industrial, scientific and medical) band with many existing wireless standards such as ZigBee, Bluetooth, Wi-Fi and LR-WPAN.

The transceiver must be low power, low cost, reliable and able to operate under varied environmental conditions regardless of RF interference and noise. The RF front-end building blocks must meet drastic power constraints. In fact, the power consumption depends on the required signal-to-noise ratio (SNR) to transmit the data (commonly expressed as bit-error-rate BER, in digital transmission). In order to optimize the power consumption, one possibility consists in quantifying as precisely as possible the circuit noise floor. Knowing that allows to optimize significantly the over-power consumption of the building blocks.

At the heart of transceivers is an RF voltage controlled oscillator (VCO), as shown in Fig. 2. It generates signals that can be used for frequency synthesis and clock generation for

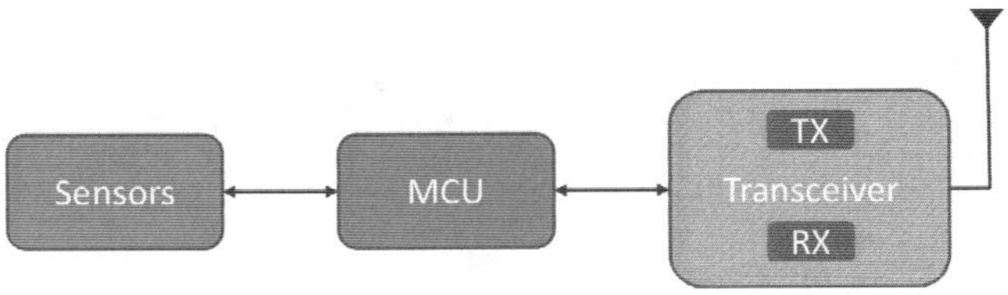

Figure 1. IoT sensor node topology.

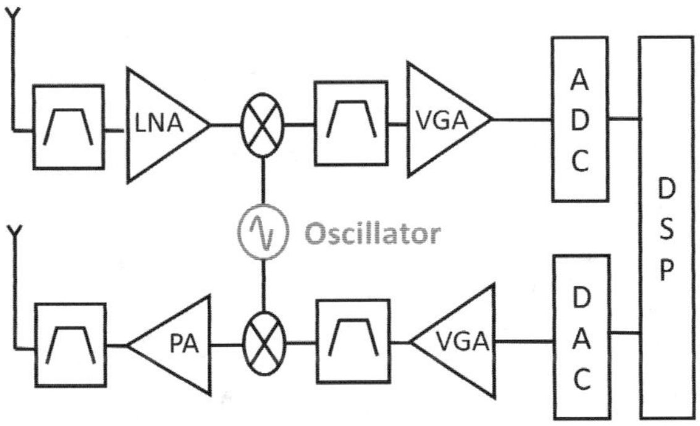

Figure 2. Generic wireless transceiver architecture.

example. VCOs can be divided into two main categories: LC-VCO and ring-VCO. Ring oscillators are characterized by a wide tuning range and low area occupation, while LC oscillators have better phase noise compared to ring oscillators due to the high Q inductors.

In this paper, a low-cost RF VCO is designed and implemented in CMOS technology. Noise characteristics, especially in terms of phase noise, are investigated and measured under different environmental conditions. The rest of the paper is organized as follows. In section II, the VCO topology is introduced and its phase noise analysis in noisy environment is presented. In section III, noise measurements are presented and analyzed. Section IV draws some conclusions.

2. CMOS VCO design

Research on wireless communications has drawn a trend towards design of highly integrated, low-cost and low-power circuits and systems. To achieve these design goals, different topologies of transceivers have been proposed in the literature [5]. Heterodyne transceivers have been widely used in second generation (2G) of wireless telephone technology thanks to its high performances in terms of selectivity and sensitivity. Intermediate frequency (IF) is used for the down-conversion procedure and leads to aliasing of the image frequency inside the useful signal band. The homodyne, architecture uses direct conversion method in order to overcome the image and integration problems leading to low cost solution. However, it achieves much lower performances than heterodyne receiver. “Low-IF” receiver allows combining the advantages of both architectures, i.e. high performances and fully integrated devices. Figure 2 shows a typical low-IF transceiver architecture. It contains many blocks such as low noise amplifier (LNA), filters, voltage gain amplifier (VGA), power amplifier (PA), mixer, converters (ADC, DAC), etc. It can be seen that the oscillator is an essential block in both sides (RX and TX) of the transceiver.

There are many VCO structures proposed in literature. The structure using RC has some restrictions such as the large layout area, noise performance, and small tuning range. LC

 DOI: 10.1051/e3sconf/20161204002

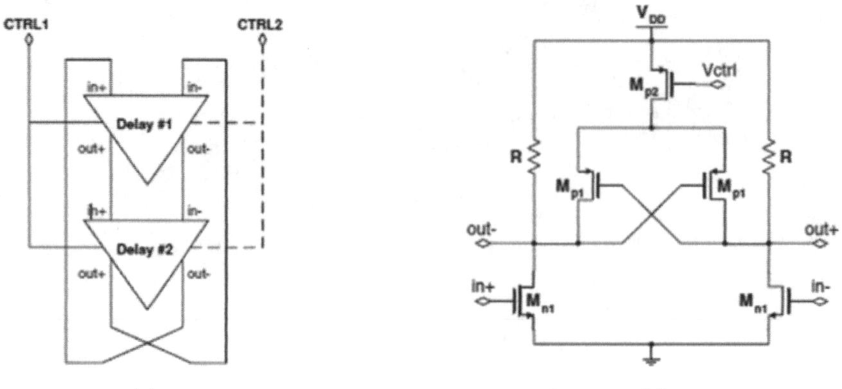

(a) (b)

Figure 3. CMOS VCO: (a) schematic with two delay stages and (b) implementation of one cell.

structure based oscillators allow very low phase noise and good spectral purity of the output frequency, but suffer of low tuning range and the difficulty to integrate the inductor part. Ring oscillators have become essential building block in VCO's [6] as their achievable performance can fit number of telecommunication standards like Bluetooth or IEEE 802.15.4 WPAN applications with relatively low area occupation and low consumption. The schematic of the well-known ring oscillator is illustrated in Fig. 3a. At first, we do not consider the dashed line corresponding to CTRL2.

A VCO, initially proposed in [7] is used to implement the oscillator function. The schematic principle of the VCO, given in Fig. 3b, uses NMOS differential pair (M_{n1}) as input stage.

It is connected to an active load formed by a cross coupled PMOS transistor (M_{p1}) which plays the role of a negative resistance, itself connected to the control transistor (M_{p2}). They are put in parallel with polysilicon resistors which are directly connected to the supply voltage (V_{DD}). If the gate voltage of M_{p2} rises, the injected current decreases and reduces in the same way the value of the negative resistance. Thus, the total load resistance of the pair increases and leads to a reduction of the oscillation frequency. At first, we do not consider the dashed line corresponding to CTRL2 (used to control the frequency temperature drift). The centre frequency of the VCO is controlled by the voltage CTRL1, as illustrated in the schematic of the implemented circuit in Fig. 4.

Only two cross-connected delay cells are used to obtain oscillations. That permits to keep power consumption low, low silicon area and reduced phase noise. In order to simplify the expression, let us note $R_T = r_{dsn1}//r_{dsp1}//R$ and $C_T = C_{dbn1} + C_{gsp1} + C_{dbp1} + C_{gdn1} + C_L$, with C_L as the load capacitance. Thus, the differential transfer function can be deduced as

$$A(s) = \frac{V_{out}}{V_{in}} = \frac{g_{mn1} R_T \left(1 + s\frac{C_{gdn1}}{g_{mn1}}\right)}{\left(1 - R_T g_{mp1}\right)\left(1 + s\frac{R_T C_T}{1 - R_T g_{mp1}}\right)} \tag{1}$$

from which, f_0 can be deduced according to Barkhausen criteria, such as $|A(j\omega_0)| = 1$, and is expressed as

$$f_0 = \frac{1}{2\pi}\sqrt{\frac{(g_{mn1})^2 - \left(\frac{1}{R} - g_{mp1}\right)^2}{C_T^2 - C_{gdn1}{}^2}}. \tag{2}$$

 DOI: 10.1051/e3sconf/20161204002

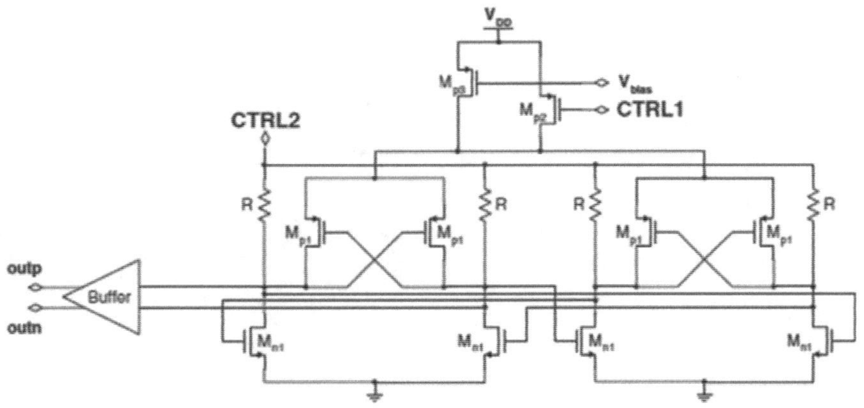

Figure 4. Schematic of the implemented VCO.

Simulations have been done using *Cadence*® tool and *Spectre-RF* simulator. A total frequency range of 453 MHz around 2.45 GHz is obtained for a control voltage extending from 0 to 2.5V with a phase noise of −102 dBc/Hz at 1 MHz offset.

3. Phase noise theory

The phase noise is considered as one of the most important factors which limit the quality of communication systems. In most applications, it is necessary that the phase noise of oscillators be kept at sufficiently low level. For an ideal oscillator operating at f_0, the spectrum assumes the shape of an impulse. Nevertheless, practical oscillators, like any other RF components, are non-ideal, always have intrinsic noise that modulates the signal and causes fluctuations in its amplitude and phase, leading to imperfect spectral purity. The oscillator's frequency spectrum exhibits sidebands around the carrier f_0 as shown in Fig. 5. As can be seen, the phase noise effectively causes the frequency of the oscillator fluctuate around f_0. Therefore, when an oscillator is used in systems, its signal contains not only the real RF signal but also unwanted RF noise signals causing degradation in the performance of corresponding systems. For instance, an oscillator with high phase noise produces large unwanted sideband noise signals, possibly causing radiation of unwanted signals for transmitters, unwanted mixing products in receivers, degradation of the signal-to-noise ratio (SNR) of receivers, etc.

We consider a general signal spectrum of oscillators as shown in Fig. 5. The phase noise at frequency f_m offset from the carrier's frequency f_0 is defined as the ratio of a single-sideband noise power (P_{noise}) in a 1-Hz bandwidth at f_m to the carrier's total power (P_{signal}) [5, 8].

The Hajimiri and Lee model [9] is generally used to calculate the theoretical phase noise of oscillators. For a ring oscillator, the pulse sensitivity root mean square function, noted Γ_{rms}, can be calculated with the following expression:

$$\Gamma_{rms} = \sqrt{\frac{2\pi^2}{3\alpha^3}\frac{1}{N^{1.5}}} \tag{3}$$

where N is the cell number that composes the VCO, α is a proportionality constant depending to the used structure. In our case $\alpha = 0.9$, so that $\Gamma_{rms} = 3/N^{1.5}$. The phase noise is then

DOI: 10.1051/e3sconf/20161204002

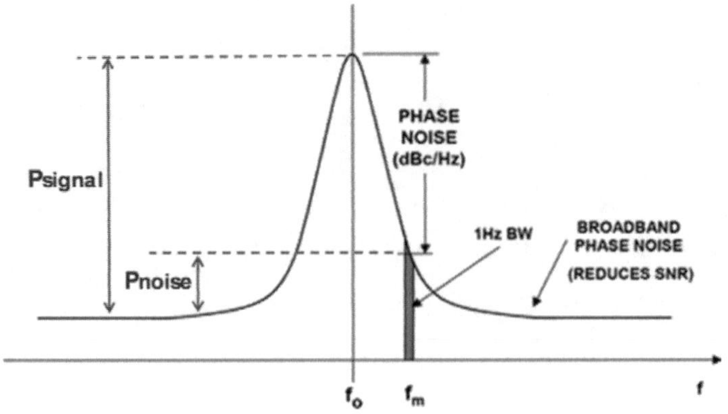

Figure 5. Oscillator's frequency spectrum showing the effect of phase noise on frequency.

calculated as

$$L(f_m) = \frac{{\Gamma_{rms}}^2}{8\pi^2 {f_m}^2} \frac{\left(\overline{i_n^2/\Delta f}\right)_{total}}{q_{max}^2} \cdot 2N. \tag{4}$$

The spectral density of noise current sources inside the circuit can be written as

$$\left(\frac{\overline{i_n^2}}{\Delta f^2}\right)_{MOS} = 4kT\gamma g_m, \quad \left(\frac{\overline{i_n^2}}{\Delta f^2}\right)_R = \frac{4kT}{R} \tag{5}$$

where γ is the channel length related coefficient of the used transistor, that, for short channel MOS in the saturation region is typically equal to 2, k is the Boltzmann constant, T is the absolute temperature, and g_m is the transconductance. We can calculate the total noise spectral density by summing that of each element such as

$$\left(\frac{\overline{i_n^2}}{\Delta f^2}\right)_{total} = \left(\frac{\overline{i_n^2}}{\Delta f^2}\right)_{n1} + \left(\frac{\overline{i_n^2}}{\Delta f^2}\right)_{p1} + \left(\frac{\overline{i_n^2}}{\Delta f^2}\right)_R + \frac{1}{4}\left(\left(\frac{\overline{i_n^2}}{\Delta f^2}\right)_{p2} + \left(\frac{\overline{i_n^2}}{\Delta f^2}\right)_{p3}\right). \tag{6}$$

Thus a theoretical phase noise of -101 dBc/Hz is obtained at 1 MHz offset from the carrier. This value is very closed to the simulation result of -102 dBc/Hz.

4. Noise measurements

A two-stage ring oscillator working around 2.45 GHz has been fabricated in a 0.28 CMOS technology. The photograph of the circuit test chip is depicted in Fig. 6a. The VCO occupies a low area of 110 μm $\times$ 110 μm.

The measurement setup of on-chip RF circuits is very complex, especially if noise requirements are stringent (case of RF oscillators). The used probe station is shown in Fig. 6b. First, DC and RF probing are positioned into the PADs on the wafer. Then, they are connected with RF cables to the spectrum analyzer. Special attention must be paid to the probing 50Ω calibration to meet measuring instruments needs. Besides, resolution bandwidth (RBW) of the spectrum analyzer should be calibrated, since it can be important versus the noise bandwidth. Noise floor of the setup measurement need to be calibrated as well.

Measurements were performed at nominal operating conditions: room temperature, supply voltage of 2.5V and a control voltage of 1V. In these conditions, noise and

 DOI: 10.1051/e3sconf/20161204002

(a) (b)

Figure 6. Photograph of the circuit test chip (b) on-chip measurement setup.

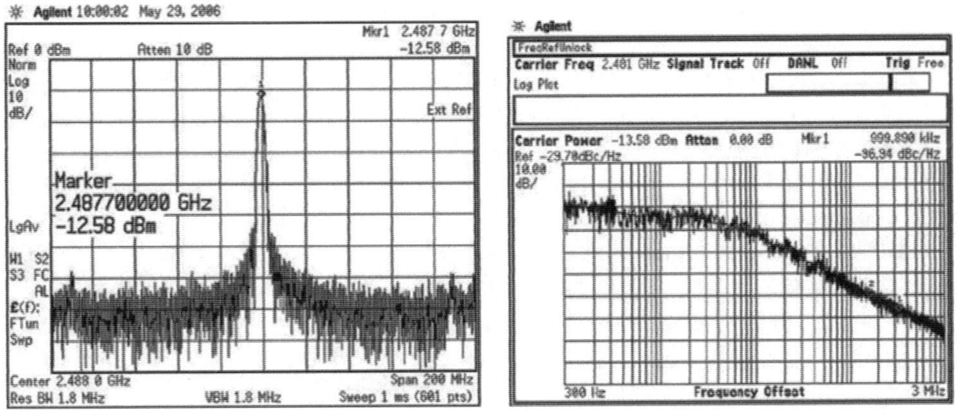

(a) (b)

Figure 7. VCO measured (a) frequency spectrum and (b) phase noise profile.

interferences are all around the measuring station, and will certainly degrade the noise floor characteristic of the VCO.

In one side, the output frequency spectrum measured from one circuit sample is depicted in Fig. 7a. It exhibits a signal power of −12.58dBm at 2.48 GHz. In the other side, the phase noise of the total circuit profile is depicted in Fig. 7b and shows a −96.9 dBc/Hz at 1 MHz offset from the carrier. A slight degradation of the phase noise is then observed during the measurement under noisy environment. RF circuit characterization such as noise measurement is usually done with anechoic chamber. Nevertheless, the calibration of the chamber requires high cost equipment like vector network analyzers (VNA) and "Termination-VSWR" method to measure the reflectivity of an absorber wall in an anechoic chamber. Moreover it requires sophisticated mathematical treatments to be made on the measured signals such as Fourier transform, allowing calculation of echoes in the time domain, filtering and removal of unwanted echoes (time-gating capability), and returning back to the frequency domain with an improved measurement accuracy.

LSSB facilities with well-known noise characteristics and calibrated environment are a good opportunity to perform VCO phase noise measurement as a future work. One must point out the VCO's intrinsic noise floor characterization in order to investigate the phase noise profiles in different environments.

5. Conclusion

VCO's are an integral part of building block for *IoT* wireless transceivers. It was given in this paper that random fluctuations in the output frequency of VCO's, expressed in phase noise, have a direct impact on the timing accuracy where phase alignment is required and on the signal-to-noise ratio where frequency translation is performed. Then, phase noise in RF oscillators has been investigated. A 2.45 GHz multi-controlled ring oscillator has been implemented with full CMOS process.

Measurement setup and obtained results in a nominal condition (a room environment with noise and interferences around the measuring station) have been presented. It was noted that noise floor needs to be performed in low-noise environment. Thus, VCO noise characterization will be performed in the near future in LSBB facilities.

References

[1] M. Lieshout, et al., "Converging Applications enabling the Information Society", Report Community research, *Future Technologies Division of VDI Technologiezentrum GmbH* (2010).
[2] Gartner, "IoT rapid prototyping to proof your business case", Gartner Symposium/ITxpo, Barcelona, Spain (2015).
[3] IEEE Standard for Local and Metropolitan Area Network – Part 15.4: Low Rate Wireless Personal Area Network (LR-WPAN), IEEE Standard 802.15.4-2011 (2011).
[4] Bluetooth Specification, Core package v4.0, Bluetooth-SIG, Kirkland, WA, USA (2010).
[5] B. Razavi, *RF Microelectronics*, Prentice Hall 2nd Edition (2011).
[6] A. Hajimiri, S. Lymotyrakis, and T.H. Lee, "Jitter and phase noise in ring oscillator." IEEE Journal of Solid-State Circuits, **34**, pp. 790–804 (1999).
[7] W. Rahajandraibe, et al., "Temperature Compensated 2.45 GHz Ring Oscillator with Double Frequency Control," in Proc of IEEE Radio Frequency Integrated Circuits Symposium (RFIC'07), Honolulu, Hawaii, pp. 409–412 (2007).
[8] C. Nguyen, *Radio frequency integrated circuit engineering*, Wiley series in Microwave and optical engineering, John Wiley & Sons (2015).
[9] A. Hajimiri and T.H. Lee, "A general theory of phase noise in electrical oscillators." IEEE Journal of Solid-State Circuits, **33**, no. 2, pp. 177–194 (1998).

 DOI: 10.1051/e3sconf/20161204003

Statistical investigations of an ENIG Nickel film morphology by Atomic Force Microscopy

Rosine Coq Germanicus[1,3], Philippe Leclère[2], Eric Hug[3], Florent Lallemand[4], and Philippe Descamps[1]

[1] Normandie Université ENSICAEN / CRISMAT / UMR 6508, 6 boulevard Maréchal Juin, 14050 Caen Cedex 04, France
[2] Service de Chimie des Matériaux Nouveaux, CIRMAP, Université de Mons, Belgium
[3] CRISMAT, Normandie Université, UMR CNRS 6508, 14000 Caen, France
[4] IPDiA, 2 rue de la girafe, 14000 Caen, France

Abstract. The morphology of a Nickel layer grown by an Electroless Nickel Immersion Gold (ENIG) technique used for microelectronics interconnections is determined by Atomic Force Microscopy (AFM) investigations. The root mean square (rms) roughness, determined over a scanned area is a function of the AFM scanned area size. In this work, we propose to consider the dynamic scale theory and the power spectrum density (PSD) analysis in order to perform a comprehensive determination of the surface properties of the ENIG nickel layer. Results highlight the existence of a first regime with a roughness exponent of 0.95 and a fractal dimension (DF) of the nickel film about 2.05. This case study is presented in order to propose further investigations. In fact, same experimental procedure should be performed in a magnetic shielded zone where a very low noise level is available such as the Low-Noise Underground Laboratory (LSBB) of Rustrel (France).

1. Introduction

In order to explore optical, mechanical, electrical and chemical properties of surfaces, quantitative investigations of surface morphology are important requirements to perform comprehensive studies. At the nanoscopic scale, Atomic Force Microscopy (AFM) appears to be a useful research tool in a wide range of different scientific fields such as materials, physics, chemistry, geology, and biology [1–4]. The AFM probe consists in a cantilever equipped with a sharp tip with a diameter of 5–30 nm. Based on the interactions between this tip and the probed surface, the deflection of the cantilever is measured and controlled via an optical system. While the tip scans the surface, a feedback loop is used to determine the topography when the system is maintaining a constant deflection value. In the last 30 years, AFM instruments have been developed to easily obtain 3D-mapping surface topographies and nanostructures with stable and high signal-to-noise ratio regarding the X, Y and Z resolution. The vertical resolution (Z), called sensitivity, is determined by the thermal and acoustic noise of the cantilever. For a free rectangular cantilever with a spring constant k of 1 N/m, when measurements are performed at room temperature in standard laboratory environment, the thermal noise is near 0.074 nm [5]. The lateral resolution (X,Y) depends on the tip apex geometry and on the tip convolution with the surface.

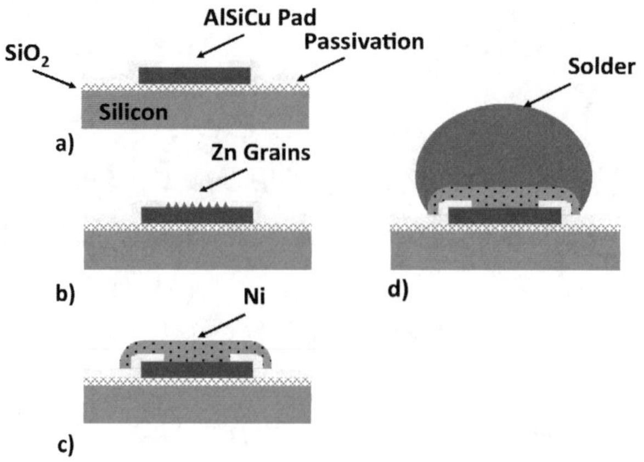

Figure 1. Schematic of the ENIG process for interconnections with Solder a) AlSiCu Pad, b) Zincation pre-plating, c) ENIG, d) Solder ball.

From topographic images, for a first order statistics, the morphology of the sample is often characterized by the Root Mean Square (RMS) roughness. This statistic parameter is determined over the fixed scanned area. For more complex film morphologies, this unique value cannot be sufficient. For deeper studies, dynamic scale analysis and Power Spectral Density (PSD) were developed [6] to describe more accurately the surface properties. The analysis is based on the self-similarity or self-affinity properties [7]. In this study, we propose to use the dynamic scale analysis of a nickel layer surface grown in a microelectronic clean-room as an UBM (Under Bump Metallization) for interconnections.

2. Experimental details

In order to realize interconnections for highly integrated circuits, the Electroless Nickel Immersion Gold (ENIG) technique is widely used in microelectronic before components assembly [8]. The goal of this process consists in obtaining a Nickel film on interconnections pads, with a high uniformity and excellent conductivity. The main steps of ENIG process is depicted on Fig. 1. The surface morphology of the nickel film is a critical parameter for Back End of Line (BeOL) [9]. In the studied sample, the metal finishing is an aluminium alloy (AlSiCu) deposited on bond pads (Fig. 1a). The diameter of the pad is fixed at 50 μm. In a second step, a pre-plating process called "zincation" activates the metallic surface (Fig. 1b). Then two metallic layers: a thick nickel coated with a thin film of gold are grown on the pads (Fig. 1c). This UBM represents the interfacial layer between the metal finishing and the solder balls (Fig. 1d) used for connecting the integrated circuit (IC) with the package or with an over IC. The electroless time is controlled in order to create a nominal nickel thickness of typically 5 μm on the pad.

In this study, the ENIG process was stopped after the nickel growth. As a consequence, the thin gold film that is typically coated on the nickel is not investigated here. The microstructure of the deposit is examined by Scanning Electron Microscopy (SEM) and AFM. The AFM images are recorded in Tapping mode using a D3100 AFM (Bruker Nano Inc. Santa Babara, CA, USA) under standard atmospheric conditions. Commercial tips with a radius of about 8 nm were used. AFM images of 512 $\times$ 512 pixels were obtained. The scan rate was fixed at 0.5 Hz. Then, from acquisitions, film morphology and roughness were determined by using the Nanoscope software (Nanoscope 1.5).

 DOI: 10.1051/e3sconf/20161204003

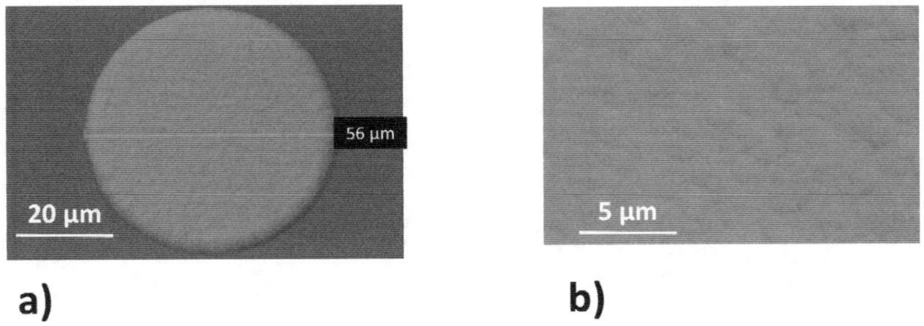

Figure 2. SEM images of the Nickel surface on pads.

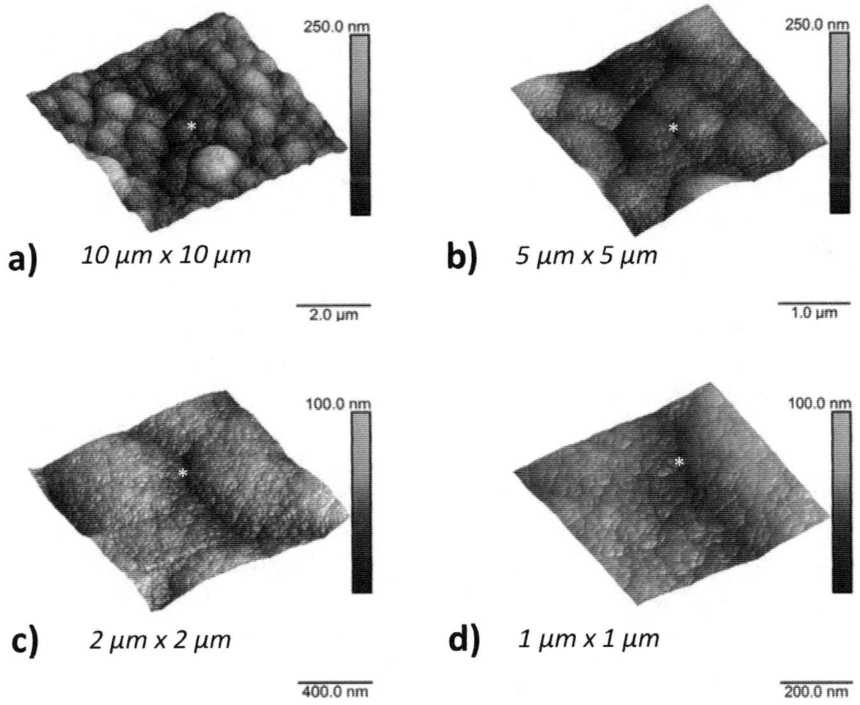

Figure 3. TM AFM topographic images of the nickel surface after electroless, for 4 scan sizes on the same area from 10 μm × 10 μm to 1 μm × 1 μm.

3. Results and discussion

Figure 2 shows the SEM images of the pad surface after ENIG Nickel growth. For an ENIG nickel film nominal thickness of 5 μm, the pad presents a total diameter of 56 μm after the electroless process. As shown in Fig. 2b, at higher magnification, the nickel layer consists in a nodular structure. In order to study and to determine the nickel film morphology, AFM measurements were performed.

AFM topographies at the same location for four scan sizes, from 10 μm × 10 μm to 1 μm × 1 μm are presented in Fig. 3. From the large scan sizes, we can appreciate the 3D view of the nodular structure of the nickel film. By decreasing the scan size Fig. 3c and d, the local structure of the film is highlighted.

The analyzed nickel surface exhibits granular properties, due to the surface anisotropic electroless process but also to the morphology of the under-layer (zincation pre-plating).

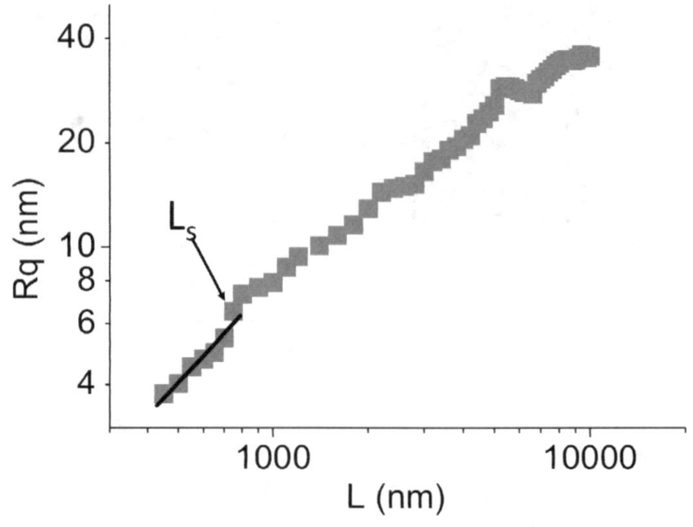

Figure 4. Plot of the roughness as a function of the scan size L.

Typically, at a first order statistic, the film microstructure is characterized by the Root Mean Square (RMS) roughness, also called Rq, which represents the standard deviation of surface heights. The average roughness Ra which indicates the height deviation is also a parameter considered for roughness analysis. The evolution of both Ra and Rq parameters determined inside the fixed 3D scanned area as a function of the surface size and also depending of the image pixel surface. In order to determine the film morphology, the dynamic scale theory [10] is used to define the RMS roughness law as a function as the image size. The plot of the roughness against the scanning size is shown in Fig. 4, this curve is derived out from the AFM topographic image (10 μm × 10 μm) presented in Fig. 3. In this case the analysis is performed with a constant surface per pixel. The graphic highlights the existence of a first scale regime. This first regime represents the film properties within an individual grain. The correlation length L_s is estimated to be about 750 nm. This value is in agreement with the average diameter of the grain. The roughness exponent define by $Rq(L) \sim L^{\alpha}$ is equal to 0.95. The fractal dimension of the film, is then determined at 2.05 (with $DF = 3 - \alpha$). From [11], the analysed surface can be considered that a relative smooth surface.

We also applied the Power Spectral Density (PSD) analysis. PSD relates the Fourier transform of the AFM image in spatial domain [12]. PSD functions are calculated from AFM measurements obtained for the four scan sizes from 1 μm × 1 μm to 10 μm × 10 μm. This analysis provides also information on the lateral roughness distribution. The PSD results versus spatial frequency are plotted on log-log scale in Fig. 5. For high spatial frequencies, the slope of the log-log plot H provides the fractal dimension, deduced from $DF = 4 - H/2$. The calculation results $DF = 2.06$ in agreement with the dynamic scale analysis.

Calculated in 3-D, this DF value seems indicated that the deposition mechanism is a diffusion-limited growth [13]. The fractal dimension analysis is performed to evaluate quantitatively a single surface parameter q for the studied surface. Our motivation is to propose the same experimental analysis in the Low-Noise Underground Laboratory (LSBB) in order to determine the impact of the noise in topography AFM analysis. We propose to perform AFM measurements for a similar surface in the LSBB in order to probe surface properties in static or dynamic modes. The goal should be to compare ground and underground analysis and quantify the gain in terms of noises in the LSBB environment.

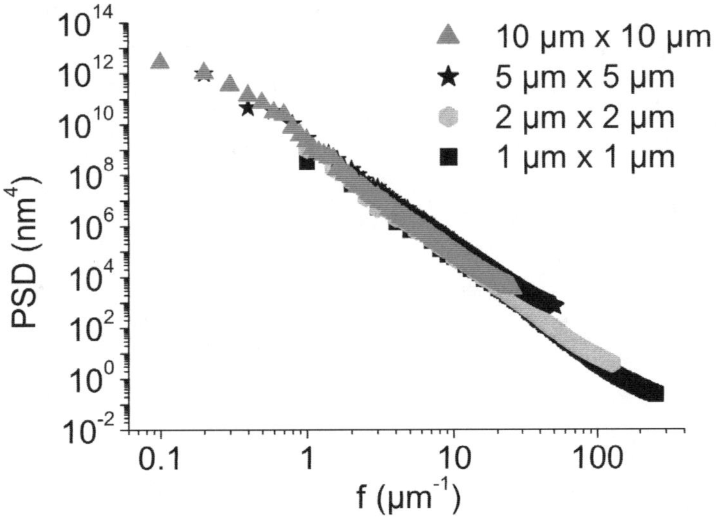

Figure 5. PSD results for the four scan sizes.

4. Conclusion

In this study, we have explored a comprehensive determination of the surface morphology of an ENIG nickel layer grown for microelectronics BEoL process. From AFM acquisitions, the dynamic scale theory and the PSD analysis were applied to determine the roughness exponent and the fractal dimension. The results indicate a roughness exponent of 0.95 and a fractal dimension of the nickel film of about 2.05 with a good correlation between the two techniques. Nevertheless, since experiments presented in this paper are performed in a standard conditions, potential electromagnetic noise might have impacted the recorded data. Thus in order to increase the accuracy on this methodology for fractal surface analysis, further investigations should be performed in a magnetic shielded zone where a very low noise level is available such as the LSBB of Rustrel (France). These very specific environment conditions, where noisy sources are reduced, should be able to provide a far better sensitivity.

References

[1] P. Eyben, W. Vandervorst, D. Alvarez, M. Xu, and M. Fouchier, *Scanning Probe Microscopy*, Springer New York, **31** (2007).
[2] R. Coq Germanicus, E. Picard, B. Domenges, K. Danilo and R. Rogel, Applied Surf. Sci. **253**, 6006 (2007).
[3] L. Leyssenne, S. Wane, S. Massenot, D. Bajon, R. Coq-Germanicus, P. Descamps and G. Audoit, ICEAA 1032 (2013).
[4] Ph. Leclère, P. Viville, M. Jeusette, J.P. Aimé and R. Lazzaroni, *In Scanning Probe Microscopies: beyond imaging,* 175 (Wiley-VCH, 2006).
[5] Y. Gan, Surface Sci. Rep. **64,** 99 (2009).
[6] L. Vázquez, R.C. Salvarezza, P. Herrasti, P. Ocón, J.M. Vara and A.J. Arvia, Applied Surf. Sci. **70** (1993).
[7] R.R.L. De Oliveira, D.A.C. Albuquerque, F.L. Leite, F.M. Yamaji, and T.G.S. Cruz, *Measurement of the nanoscale roughness by atomic force microscopy: basic principles and applications* INTECH (2012).

[8] B.K. Kim, S.J. Lee, J.Y. Kim, K.Y. Ji, Y.J. Yoon, M.Y. Kim, S.H. Park, and J.S. Yoo, Journal of Electr. Mat., **37**(4), 527 (2008).

[9] J. Wojewoda-Budka, Z. Huber, L. Litynska-Dobrzynska, N. Sobczak and P. Zieba, Mat. Chem. and Phys. **139**, 276 (2013).

[10] N.C. De Souza, J.R Silva, M.A. Pereira-da-Silva, M. Raposo, R.M. Faria, J.A. Giacometti, Journal of Nanosci. and Nanotech, **4**(5), 548 (2004).

[11] A.P. Pentland, IEEE transactions on pattern analysis and machine intelligence, **6**, 661 (1984).

[12] F. Biscarini, P. Samori, A. Lauria, P. Ostoja, R. Zamboni, C. Taliani, P. Viville, R. Lazzaroni and J.L. Bredas, Thin Solid Films, **284**, 439 (1996).

[13] T. Vicsek, *Fractal growth phenomena* **2**, Singapore: World scientific (1992).

Polarisation properties of [SQUID]^2 horizontal components at the LSBB (*Laboratoire Souterrain à Bas Bruit*)

Christian Kwisanga[a]

Department of Physics, College of Science and Technology, University of Rwanda, Rwanda

Abstract. Polarisation properties of the geomagnetic signal are computed using the coherence matrix of horizontal components (EW and NS) of the [SQUID]^2 datasets, in a relatively small bandwidth. Wavefront ellipticity, signal-to-noise ratio (*snr*) and wavefront arrival angle of the magnetic quasi-monochromatic waves are determined. From the variation of ellipticity extremum position in the vicinity of 8 Hz, the temporal variation of the peak frequency is traced for the LSBB and two Northern American stations distant from the LSBB by $\sim$8,000 km. The spectra of the peak frequency variation display the daily, half daily and third-daily harmonics at all stations, which are characteristic of the first Schumann resonance. Ellipticity spectrograms also unveil a type of chirping local nighttime resonances, known as ionospheric Alfvén resonances (IAR), observed at all stations. Thanks to the *snr* spectrograms, components of the signal which are local to the LSBB station are cancelled from the output, particularly the 50-Hz power grid signal which is minimised in the *snr* spectra.

1. Introduction

At the extremely low frequency (ELF [3–3000 Hz]) range, the main source of natural electromagnetic energy in the Earth-ionospheric cavity is due to thunderstorm activity and the cavity that acts as a resonator for specific modes called Schumann resonances (SR) [1]. The sources of electromagnetic pulse excitation is random in time and space around the globe, with hotspots in continental tropical areas. The interference between electromagnetic pulses generated worldwide forms SR with eigenfrequencies at 7.8, 14, 20 Hz [2]. The random nature of generation and phase distribution of pulses form at a ground station a superposition of fully polarised and depolarised waves [1].

The polarisation study of geomagnetic components in the ELF range is an important tool to assess whether electromagnetic SR measured at an observatory are of global nature. Coherence is the measure of the degree of polarisation of electromagnetic waves. It has two aspects: Spatial and temporal coherence. Spatial coherence determines the correlation between waves at different points in space and temporal coherence defines their correlation between wavefronts at different instants in time. Here, time coherence is evaluated by determining the rate of variation of the amplitude and phase of the electromagnetic wavefront. Polarisation properties of the magnetic wave are computed using quasi-monochromatic wave theory in the frequency domain [3]. The method is based on the auto-correlation and cross-correlation of components of the power wavefield called the coherence matrix (CM).

[a] e-mail: kwistiand@gmail.com

In this paper, we attempt to establish a correlation between magnetic datasets measured at the LSBB using the Superconductive Quantum Interference Device Qualified for Ionospheric Detection ([SQUID]^2) and the Canadian Array for Realtime Investigations of Magnetic Activity (CARISMA) network using the high permittivity Induction Coil Magnetometers (ICM).

2. Computation of polarisation properties

The non-polarised and polarised components of the horizontal geomagnetic field (H) can be resolved by computing eigenvalues and eigenvectors of the CM, considering H components as quasi-monochromatic waves [3]. The two components are

$$H_X = A_1(t)\exp[i(\bar{\omega}t + \varphi_1(t))] \tag{1}$$

$$H_Y = A_2(t)\exp[i(\bar{\omega}t + \varphi_2(t))] \tag{2}$$

where $\bar{\omega} = 2\pi\bar{\upsilon}$ is the average angular frequency of the wave packet, $A_1(t)$, $A_2(t)$ respective time-dependent amplitudes and $\varphi_1(t)$, $\varphi_2(t)$ are time-varying phases.

2.1 Coherence

The polarisation of the wavefield can be represented as the coherence matrix (CM) of the form

$$J = \begin{bmatrix} J_{XX} \; J_{XY} \\ J_{YX} \; J_{YY} \end{bmatrix} = \begin{bmatrix} < H_X H_X^* > \; < H_X H_Y^* > \\ < H_Y H_X^* > \; < H_Y H_Y^* > \end{bmatrix} \tag{3}$$

where H_X^*, H_Y^* are complex conjugates to the horizontal magnetic waves. In case the wave is fully polarised, then $\det(J) = 0$. The intensity of the total wave packet would be obtained by plugging (1) and (2) into (3)

$$I = J_{XX} + J_{YY} = A_1^2 + A_2^2. \tag{4}$$

In case the signal is not fully polarised, $\det(J) \neq 0$. The degree of polarisation of the wave is the ratio of the polarised to the total wave intensity. By solving eigenvalues of (3), the degree of polarisation (coherence) can hence be expressed as

$$P_D = \left[1 - \frac{4|J|}{(J_{XX} + J_{YY})^2}\right]^{\frac{1}{2}}. \tag{5}$$

P_D is always positive and varies from 0 to 1. $P_D = 0$ represents a completely depolarised signal and $P_D = 1$ a totally polarised signal.

2.2 Ellipticity

Ellipticity is a property that determines the type and the direction of rotation of the polarised electromagnetic wave in time. Elliptical polarisation properties are obtained from the CM as follows [4]

$$\frac{Y^2}{P_{YY}} + \frac{2Re\,P_{XY}}{P_{XX}P_{YY}} + \frac{X^2}{P_{XX}} = 1. \tag{6}$$

Equation (6) is a typical equation of an ellipse. Parameters associated with the ellipse are depicted in Fig. 1: The direction of polarisation β and the angle of inclination of the ellipse θ.

DOI: 10.1051/e3sconf/20161202002

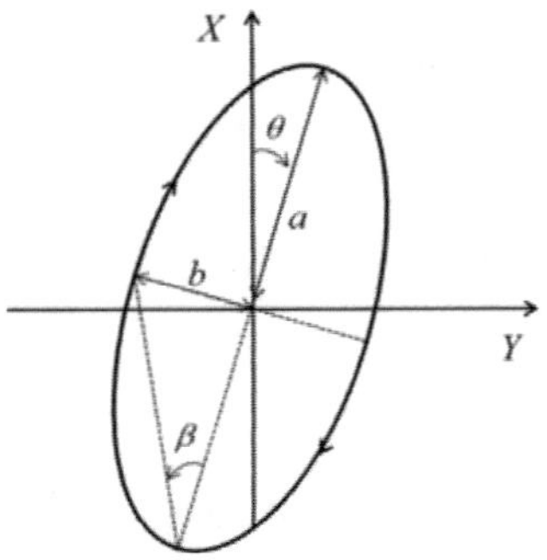

Figure 1. Sketch of the elliptical polarisation of the horizontal H, where a, b represent the semi-major and semi-minor axes of the ellipse. Axes x and y represent the WE and SN directions. θ represents the inclination angle of the major axis of the ellipse to the X axis. The direction of polarisation represented here is right handed in the direction of z [5].

β can be deduced from (6) as $\sin 2\beta = \left[i(P^*_{XY} - P_{XY})\right]\left[(P_{XX} - P_{YY})^2 + 4P^*_{XY}P_{XY}\right]^{-\frac{1}{2}}$. Ellipticity is hence

$$el = \tan\beta = \frac{b}{a}\cdot \tag{7}$$

By convention, when the angle $\beta < 0$ ($el < 0$), the polarisation is clockwise or right-handed as shown in Fig. 1, and $\beta > 0$ ($el > 0$), the polarisation is counter-clockwise or left-handed. The value of ellipticity varies as $|el| \leq 1$, with $el = 0$ describing linear polarisation and $|el| = 1$ circular polarisation.

2.3 Inclination angle of the polarisation ellipse

The inclination angle is the angle made between the semi-major axis and the X axis as shown in Fig. 1. The relationship between the inclination of the ellipse and the coherence matrix is [4]

$$\tan 2\theta = \frac{2Re P_{XY}}{P_{XX} - P_{YY}}\cdot \tag{8}$$

θ is by convention taken positive in the clockwise direction and is referred to as the arrival angle of the H wave.

3. Signal-to-noise ratio (*snr*)

The decoherence (depolarisation) coefficient can be defined in function of the polarisation (coherence) coefficient of the wave as [6]

$$D_D = 1 - P_D, \tag{9}$$

with P_D defined in (5). In the same way as P_D, $0 \leq D_D \leq 1$. By referring to the coherent component as the signal $S(f) = H(f)\cdot P_D$, and the noise $n(f) = H(f)\cdot D_D$, where $H(f)$ is the spectrum of the horizontal magnetic signal. The snr can be approximated as

$$snr(f) = \frac{S(f)}{n(f)} = \frac{P_D}{D_D}\cdot \tag{10}$$

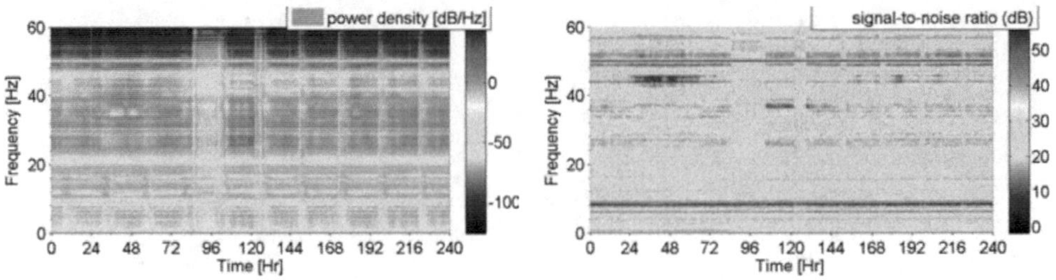

Figure 2. Ten day [SQUID]^2 H power spectral density (left) and *snr* (right graph) starting from 2008-08-22.

Table 1. Peak noise level, average frequency and peak power of the H signal from 2008-08-22 to 31.

frequency (Hz)	6	7.97	50
noise/signal ratio	0.1	0.22	0.32
Peak Power (pT^2Hz^{-1})	1	5.4	4×10^5

4. Application to LSBB datasets

Daily geomagnetic datasets sampled at 125 Hz downloaded from the LSBB server[1] are used. The time signal is subdivided in 32.8 s ($N = 4096$) segments which are transformed to the Fourier domain to obtain a spectral resolution of 3×10^{-2} Hz. The spectra are accumulated and averaged over a duration of 24 minutes using Welch's method [7]. For this study, only quiet magnetic datasets are taken into consideration. They are sampled from the minimum solar activity epoch spanning from 2007 to 2009 [5].

4.1 [SQUID]^2 Power density and signal-to-noise ratio spectra

Figure 2 shows H power spectrograms and its *snr* in a frequency range from DC to 60 Hz spanning in time from 2008-08-22 to 31. During the whole period of time, the index of magnetic disturbance is relatively low ($k_p \leq 2$), indicating a worldwide quiet magnetic activity. Maxima in the power of the signal are displayed in Table 1. The *snr* spectrogram in Fig. 2 (right) shows the lowest level at 5 dB at 50 Hz with an average level of noise which is 1/3 of the power in the coherent signal. Table 1 shows the respective peaks of the noise and peak power observed in the spectra.

The peaks of maximum power seem to coincide with the peaks of noise in the spectra as shown in Fig. 2. This might mean that these peaks are caused by an incoherent signal local to the LSBB capsule environment. In an internal report [8], it was shown that switching on the ventilation unit in the capsule caused peaks of noise as high as 30 $pT/\sqrt{Hz}$ at 10.3 Hz in all the [SQUID]^2 components.

5. [SQUID]^2 – ICM correlation

ICM polarisation properties are compared to the [SQUID]^2's in order to demonstrate the presence of a global magnetic signal component at LSBB. Two CARISMA stations distant of 1,200 km in longitude are considered as shown in Table 2. The datasets are sampled on site at 20 Hz thus display only the first SR [9]. The stations are separated from the LSBB by 8,000 km on average.

---

[1] `http://www.lsbb.eu/index.php/en/`

 DOI: 10.1051/e3sconf/20161202002

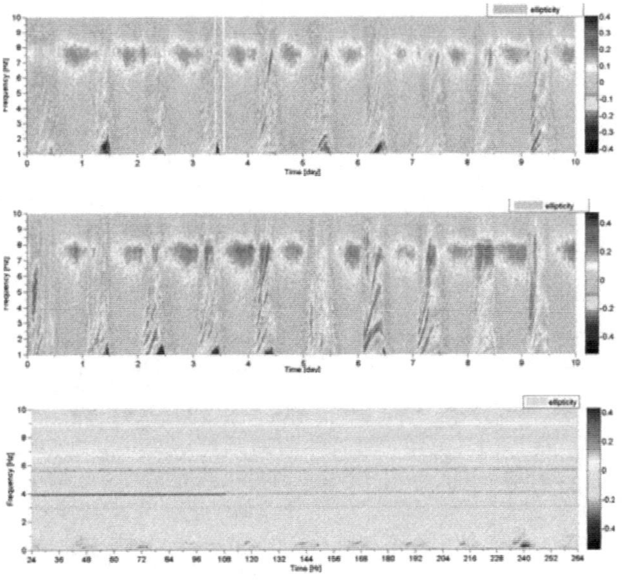

Figure 3. Ellipticity spectrograms from DC – 10 Hz spanning for 10 days from 2008-08-22, at MSTK (top), THRF (middle) and LSBB (bottom).

Table 2. ELF Magnetometer station locations and coordinates.

Site	Location, Country	Lat. North	Long. East
LSBB	Rustrel, France	44.95	87.53
MSTK	Ministik Lake, Canada	53.3	247
THRF	Thief River Falls, USA	57.82	331.49

Figure 3 shows the variation of the ellipticity at the 3 observatories. MSTK and THRF display a maximum (with $el > 0$) band in the vicinity of 8 Hz characteristic of the first SR [6], while at the LSBB the spectra display a minimum ($el < 0$). By considering $|el(f)|$, the spectra at all the stations display a peak at 8 Hz. This indicates that magnetic waves resonate at 8 Hz but are left-hand and right-hand polarised at CARISMA and LSBB respectively (see Sect. 2.2).

The spectra also display local nighttime resonant signals that occur in quiet magnetic conditions that are called ionospheric Alfvén resonances (IAR) [6]. The IAR appear as chirping resonances whose peak frequencies rise in time before they disappear at local-time dawn [6]. In H spectra, the IAR are elliptically polarised and their peak frequencies are highly dependent on the magnetic latitude. For high-latitude magnetic spectra (MSTK and THRF), they can reach frequency ranges up to the first SR (as shown in Fig. 3 top and middle graphs), while for the LSBB signal, the IAR are located in the range of quasi-DC$-$1 Hz (Fig. 3 bottom graph).

By tracking the extrema of el in the vicinity of 8 Hz, it is possible to determine the variation of the first SR (Δf_0) as a function of time (t). The maximum of the ellipticity at CARISMA station (and the minimum at LSBB) and their corresponding peak frequency are recorded in the spectra for every 12-minute. The resulting signals $el(t)$ and $\Delta f_0(t)$ are Fourier transformed, yielding $el(f)$ and $\Delta f_0(f)$ with a cutoff frequency of 60 day^{-1}. $el(f)$ and $\Delta f_0(f)$ display harmonics embedded in the time-dependent signals as shown in Fig. 4 (left). In order to study diurnal variations, the signals are filtered through a passband filter centered at the first daily harmonic ($f = 1$ day^{-1}) with respect to the bandwidth in their spectra. Figure 4 (right) shows the diurnal variations of el and Δf_0.

 DOI: 10.1051/e3sconf/20161202002

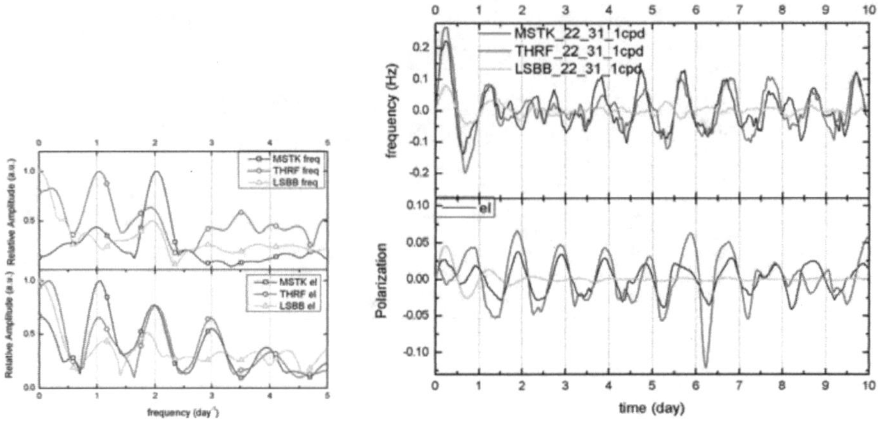

Figure 4. Diurnal variation of the frequency (top graph) and ellipticity (bottom graph) during the quiet magnetic period 2008-08-22 to 31. The spectra [left side] and the filtered signals around 1 day^{-1} [right side] are represented.

As depicted in Fig. 4, CARISMA stations show a diurnal variation in time of the ellipticity comparable to [2] and peak frequency diurnal variation that is similar to the first SR [10]. The diurnal *el* and Δf_0 variations at CARISMA stations are highly correlated. Frequency and polarisation variations at LSBB are not correlated with those at CARISMA stations and the peak ellipticity in the vicinity of 8 Hz is constant throughout the whole dataset.

6. Conclusion and recommendations

The coherence matrix has proven to be a straightforward and stable method to evaluate the degree of polarisation of geomagnetic signals at the LSBB. In the lower ELF range (1−60 Hz), the signals that are coherent can be associated with a global resonance and those uncoherent discarded as localised noise. The latter are observed at 6, 7.9 and 50 Hz in the *snr* spectra. [SQUID]^2 *H* spectra do not clearly display the first SR because the amplitude of variation of the ellipticity and peak frequency are much lower than those observed in the ICM data. Therefore, *H* polarisation properties of another (SQUID) magnetometer local to the LSBB, but not located in the capsule, should be evaluated as a comparison to the [SQUID]^2's.

The author thanks I.R. Mann, D.K. Milling and the rest of the CARISMA team for data. He is also thankful to the LSBB team and their collaborators for providing the [SQUID]^2 data.

References

[1] Nickolaenko A. P., Hayakawa M., *Resonances in the E-I Cavity* (Springer Sc., Tokyo, 2002).
[2] Sentman D.D., Radio Sci. **22**, 595–606 (1987).
[3] Born M. & Wolf E., *Principles of Optics* (Mac-Millan, New York, 1959).
[4] Fowler R.A., Kotick B.J. & Elliott R.D., Geophys. Res., **72**, 2871–2883 (1967).
[5] Kwisanga C., *SQUID Geomagnetic Signal Analysis and Modelling of SR in the E-I Cavity* (E & E Eng., Stellenbosch, 2016).

[6] Nickolaenko A.P. & Hayakawa M., *Schumann Resonances for Tyros* (Springer, Tokyo, 2014).
[7] Welch P.D., IEEE Trans. Audio & Electroac., **15**, 70–73 (1967).
[8] Bernard S., Pozzo di Borgo E., & Febvre P., LSBB Tech. Rep. 27–28 July (2010).
[9] Mann I.R., *et al.*, Space Sci. Rev., **141**, 413–451 (2008).
[10] Price C. & Melnikov A., Atmos. Sol.-Terr. Phys., **66**, 1179–1185 (2004).

 DOI: 10.1051/e3sconf/20161201001

Progress on Ultra-Wideband (UWB) Multi-Antenna radar imaging for MIGA

Matthew Yedlin[1,a], Jean-Yves Dauvignac[2], Nicolas Fortino[2], Christian Pichot[2], Raul Cova[3], David Henley[3], Kevin Hall[3], and Stéphane Gaffet[4]

[1] Department of Electrical and Computer Engineering, University of British Columbia, 2332 Main Mall Vancouver, B.C. V6T 1Z4, Canada

[2] Laboratoire d'Electronique, Antennes et Télécommunications (LEAT) Université Nice – Sophia Antipolis, CNRS Bât. Forum, Campus SophiaTech, 930 Route des Colles, 06903 Sophia Antipolis, France

[3] CREWES (Consortium for Research in Elastic Wave Exploration Seismology) Department of Geoscience University of Calgary, 2332 Main Mall Calgary, Alberta, T2N 1N4, Canada

[4] Low Background Noise Inter-disciplinary Underground Science and Technology, LSBB Underground Research Laboratory, UMS3538 AMU/CNRS/UAPV/UNS La Grande Combe, 84400 Rustrel, France and GEOAZUR, UMR 7329 UNS/CNRS/IRD/OCA, 250 rue Albert Einstein, Bât. 4, Sophia-Antipolis, 06560 Valbonne, France

Abstract. Progress on the development of the multi-channel, ground penetrating radar imaging system is presented from hardware and software perspectives. A new exponentially tapered slot antenna, with an operating bandwidth from 100 MHz to 1.5 GHz was fabricated and tested using the eight-port vector network analyzer, designed by Rhode and Schwarz Incorporated for this imaging project. An eight element antenna array mounted on two carts with automatic motor drive, was designed for optimal common midpoint (CMP) data acquisition. Data acquisition scenarios were tested using the acoustic version of the NORSAR2D seismic ray-tracing software. This package enables the synthesis and analysis of multi-channel, multi-offset data acquisitions comprising more than a hundred thousand traces. Preliminary processing is in good agreement with published bistatic ground-penetrating radar images obtained in the tunnels of the Low-noise Underground Laboratory (LSBB) at Rustrel, France.

1. Introduction

The universities of Avignon, Nice Sophia Antipolis, British Columbia and Calgary are constructing a multi-channel ground-penetrating radar(gpr) imaging system to support the **M**atter-wave laser **I**nterferometric **G**ravitation **A**ntenna (MIGA) project. This system is based on ANR Maxwell, a prototype funded by the ANR. Our research focuses on the design and construction of an 8 element ultra-wideband antenna array. Preliminary ray-tracing models of data to be collected by this new system match the images obtained by Sénéchal et al. [1].

[a] Corresponding author: matty@ece.ubc.ca

 DOI: 10.1051/e3sconf/20161201001

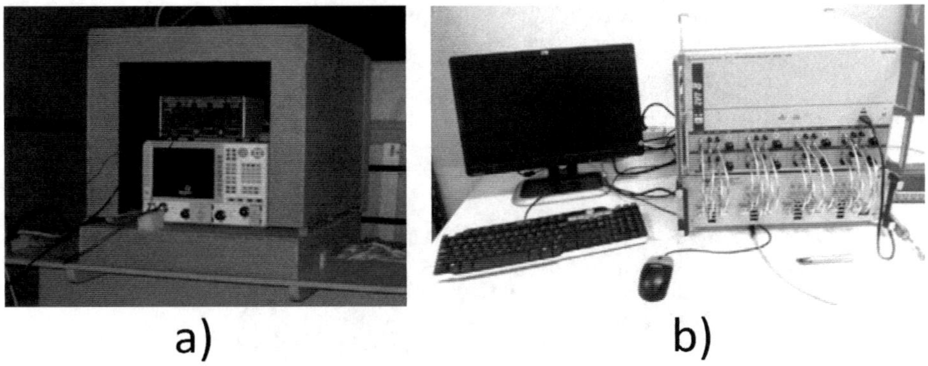

Figure 1. a) ANR MAXWELL VNA. b) Project MIGA Rhode and Schwarz VNA.

2. Imaging system progress – hardware

2.1 Antenna development

The proposed ultra-wideband (UWB) microwave imaging system consists of a customized Vector Network Analyzer (VNA) connected to eight specifically designed antennas. These antennas are integrated in an autonomous shielded carriage. The previous ANR MAXWELL gpr imaging project [2, 3] used an external amplifier as shown in Fig. 1a. The new frequency range of this UWB ground-penetrating radar extends from 100 MHz to 1.5 GHz to obtain a good trade-off between penetration depth, spatial resolution, and antenna size. This UWB system is designed to provide fast, multi-static, multipolarization, wide offset measurement capabilities with a high dynamic range and a better timing resolution than narrow band systems.

A large dynamic range, 8-port VNA, shown in Fig. 1b, connected to additional customized amplifiers, is used for recording the reflected electromagnetic waves received by the antennas. Our equipment offers high dynamic range measurement (typically 130 dB) to monitor the water content of the unsaturated area of the Vaucluse Karst aquifer under the LSBB site, at decametric depth. This new acquisition system includes 8 high gain integrated amplifiers that are specifically shielded to avoid spurious radiation which was noticeable in the previous MAXWELL system.

A new ultra wide band printed antenna was specifically designed to fit the project requirements [4, 5]. Classical Tapered Slot Antennas (TSA) have a low profile, a wide bandwidth, and superior radiation characteristics. However, the TSA size limits their integration capabilities in the antenna array carriage (size must be kept below 75 cm, which is $\lambda/4$ at 100 MHz compared to the typical $\lambda/2$). We used new loading techniques to place the CMS resistive loads to achieve this design goal without modifying antenna performance. Insertion of loads, commonly used in radar applications, increases antenna bandwidth and minimizes return losses. This technique, however, usually degrades antenna efficiency and as a result, its gain. For the proposed antennas, the placement of loads was carefully studied to extend bandwidth to lower frequencies without affecting antenna radiation at higher frequencies. Moreover, this modification increases antenna directivity. The advantages of having a higher directivity are twofold: a higher gain at low frequencies (the lack of efficiency is over-compensated) and a lower coupling between neighboring antennas in the array. This novel ETSA, shown in Fig. 2, demonstrates excellent impedance matching from 80 MHz to 2 GHz, a maximum gain of 9 dB, and an improved impulse response.

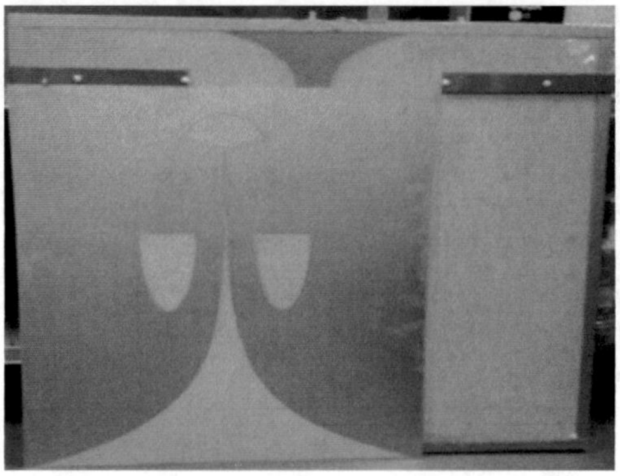

Figure 2. New miniaturized ETSA (Exponentially Tapered Slot Antenna) superimposed on original larger ETSA.The exponential taper (a horn-shape) is clearly visible at the bottom of the figure.

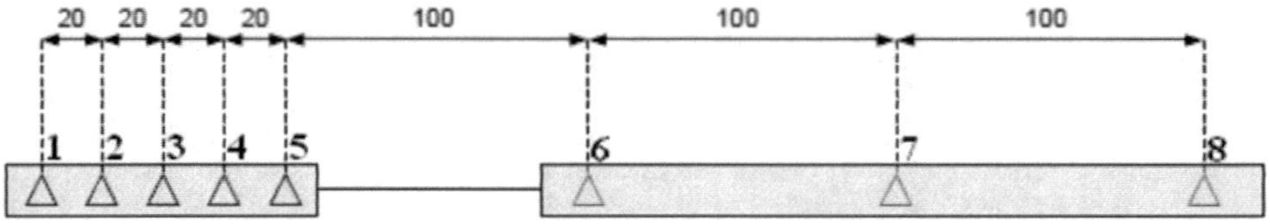

Figure 3. Schematic of antenna array with 20 and 100 cm antenna element spacing.

2.2 Antenna array

The array is separated in two carts linked together, illustrated schematically in Fig. 3. The first one is motorized and provides the traction of the array. It embeds an array of 5 shielded antennas separated by 20 cm. The second one is placed so that its first antenna is separated by 1 m from the last antenna of the tractor cart. It embeds an array of 3 shielded antennas separated by 1 m. Using this array geometry, the auscultation of a 200 m gallery with a 10 cm step generates 2500 frequency measurements for each position and each antenna transmission between the 8 antennas. The amount of collected data exceeds 19 classical CMP runs on the same zone and takes less than 17 hours with an ultra-wide bandwidth measurement. The flexibility of the instrument also offers the possibility to prioritize a fast sweep (time measurement divided by 7 for a 120 dB dynamic range) for a simple survey, or a high dynamic range (up to 140 dB) for further investigations focused on a specific zone.

3. Imaging system progress – synthetic image simulation at LSBB

3.1 Theory

This data is generated using the NORSAR seismic two-dimensional ray tracing software, in an acoustic setting. After simple linear scaling, the acoustic wave equation can be used to model TE-polarized electromagnetic waves as shown by Laurain and Lecomte [6]. They demonstrate clearly that the TE reflection coefficients correspond precisely to the acoustic reflection coefficients with appropriate conversions of impedance and velocity. For this study,

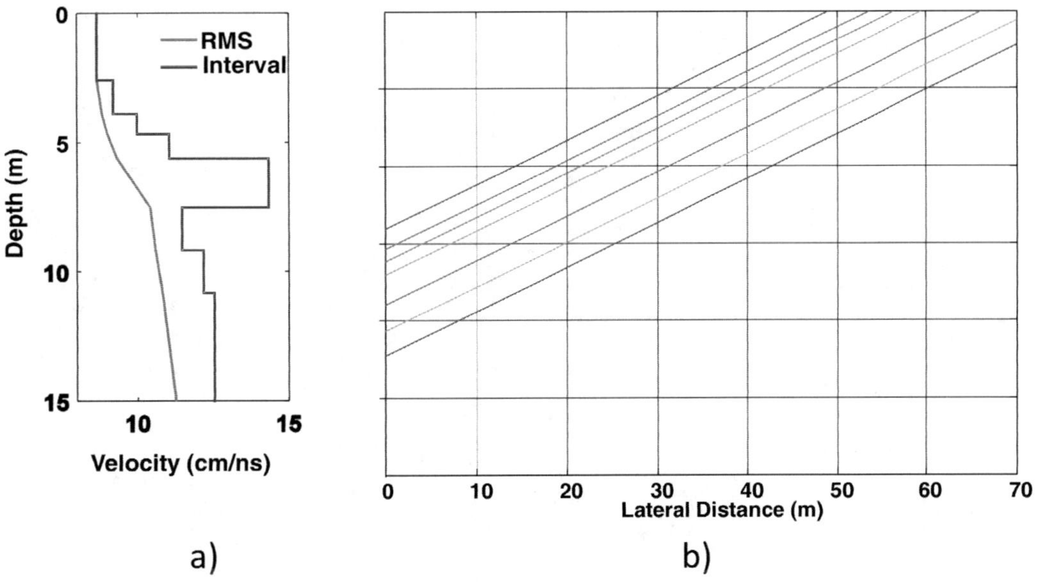

Figure 4. a) RMS and interval velocities obtained from Fig. 5a in [1]. b) Interpolated velocity model obtained from Fig. 4 in [1].

primary reflections, from layer boundaries that enclose constant parameter layers, have been utilized.

3.2 Case study

We have chosen Figs. 4, 5a and 5b from the 2013 paper by Sénéchal et al. [1] to create synthetic data of the type that would be collected by the antenna array shown in Fig. 3. Figure 4 was used to create the velocity structure model. Figures 5a and 5b were used to create the RMS and interval velocities.

Shown in Fig. 4, panel a) are the RMS velocities that were picked from the Fig. 5b in [1]. The interval velocities, shown in panel a) were obtained using the Dix equations [7], with a dip correction corresponding to a dip of 16 degrees. In Fig. 4, panel b), the structural model, corresponding to Fig. 4 in [1] is presented. The dipping layers were obtained by linear interpolation. A synthetic CMP gather, without any additive noise, at location 40 m, is presented in panel a) of Fig. 5, with associated semblance in panel b). For comparison, the CMP gather from Fig. 5a and the semblance in Fig. 5b of Sénéchal et al. [1] are presented in panels c) and d) respectively. There is very good agreement of the primary arrival hyperbolae and associated semblance panels. For example, at 60 ns, the hyperbolic caps of the synthetic and real data line up as do the maxima of the semblance panels b) and d). The alignment with the other six hyperbolae shown gives us confidence that our model-building procedure and subsequent ray tracing is adequate to model results in the future.

To validate our proposed modeling of acquired data, in Fig. 6 we present a synthetic source gather in panel a) and the complete CMP stack in panel b). The data in panel a) was created by adding a pseudo-random noise generated from a normal distribution with zero mean, resulting in an SNR of 1.3. It was computed at a source position of the antenna array at location 38 m, near 40 m, the location of the CMP used in the velocity semblance analysis. In panel b) we present the final stacked data, reflecting the original model shown in Fig. 4b.

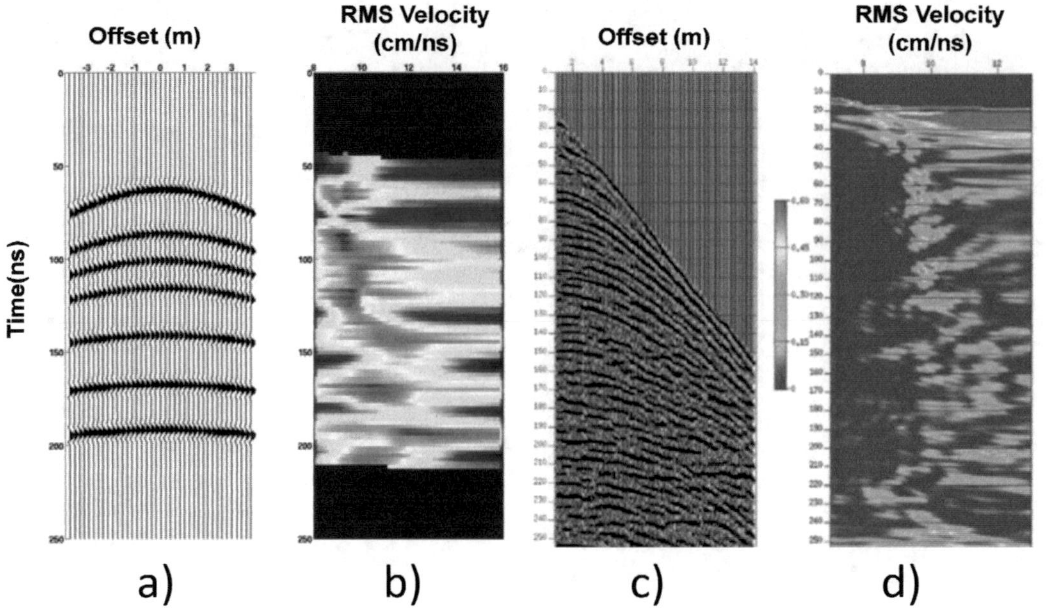

Figure 5. a) Synthetic CMP gather. b) Velocity Semblance computed from a). c) CMP recorded at LSBB [1]. d)Velocity Semblance computed from d).

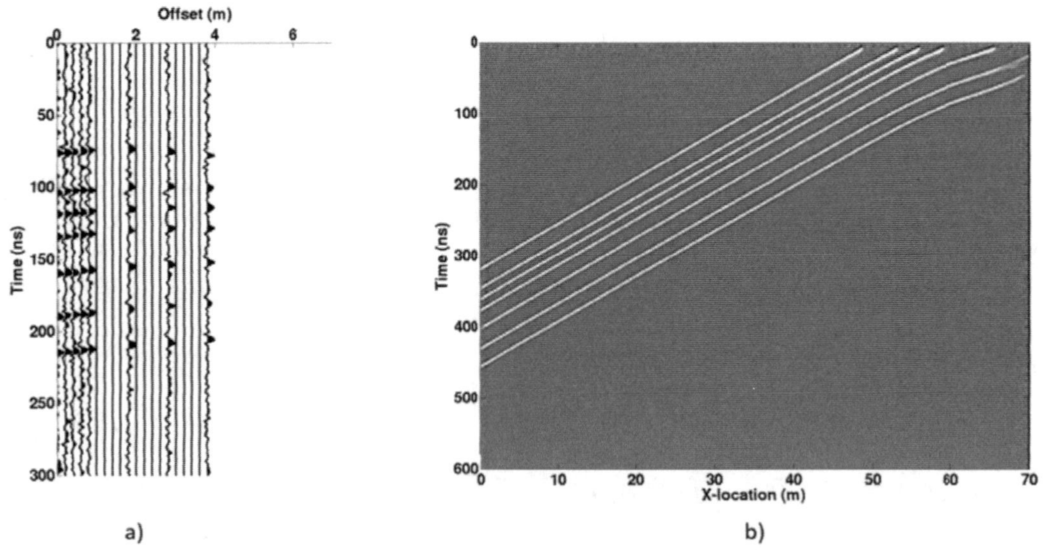

Figure 6. a) Source gather at 38 m. b) Complete stacked section for velocity model.

4. Conclusions

In our research paper, we have presented the current progress in the development of an 8 element ground penetrating radar antenna array composed of novel, compact ultra-wideband antennas. Data acquisition is achieved by using the Rhode Schwarz Incorporated ZVT vector network analyzer, with an attached eight channel shielded amplifier, designed specifically for this radar imaging project.

Concurrent progress in the development of a data simulation platform was demonstrated by using the NORSAR2D ray-tracing software to simulate data that will be recorded at LSBB. Preliminary results demonstrate the viability of the ray-tracing approach necessary to model the 128,000 traces that will be collected at LSBB. This modelling is a necessary first step in creating an inversion algorithm for the determination of subsurface permittivity.

The authors would like to acknowledge support from the ANR(France), NSERC(Canada), CREWES (Consortium at the University of Calgary) and CREWES sponsors.

References

[1] G. Sénéchal, D. Rousset, S. Gaffet, Near Surface Geophysics **11**, 283 (2013).

[2] J. Dauvignac, N. Fortino, G. Sénéchal, A. Cresp, M. Yedlin, S. Gaffet, D. Rousset, and C. Pichot, *Ultra-wideband GPR imaging of the Vaucluse Karst Aquifer*. In *AGU Fall Meeting Abstracts* (2008), Vol. 1, p. 08.

[3] M. Yedlin, G. Sénéchal, D. Rousset, N. Fortino, J.Y. Dauvignac, S. Gaffet, T. Monfret, and C. Pichot, *Comparative study using an UWB measurement system and a RAMAC GPR system for subsurface imaging of the vaucluse karst aquifer*, in *Wireless Information Technology and Systems (ICWITS), 2010 IEEE International Conference on* (IEEE, 2010), pp. 1–4.

[4] C. Diakité, N. Fortino, J.-Y. Dauvignac. *Antenne ETSA "Exponential Tapered Slot Antenna" miniature pour radar à pénétration de surface*, in *19èmes Journées Nationales Microondes (JNM 2015)* pp. Session–Antennes.

[5] G. Clementi, N. Fortino, J. Dauvignac, *A novel low profile Tapered Slot Antenna with absorbing material for radar imaging system*. in *Antennas and Propagation (EuCAP), 2013 7th European Conference on* (IEEE, 2013), pp. 2891–2895.

[6] R. Laurain, I. Lecomte, *Elastic/Electromagnetic Wave Propagation-Equivalences and 2D Modelling of GPR*, In *63rd EAGE Conference & Exhibition* (2001).

[7] Ö. Yilmaz, *Seismic data analysis*, Vol. 1 (Society of exploration geophysicists Tulsa, 2001).

Full-waveform inversion of GPR data acquired between boreholes in Rustrel carbonates

Hugo Pinard[1,a], Stéphane Garambois[1], Ludovic Métivier[1], Michel Dietrich[1], Guy Sénéchal[2], and Dominique Rousset[2]

[1] Université Grenoble Alpes, ISTerre, CNRS, 38000 Grenoble, France
[2] LFCR, Université de Pau et des Pays de l'Adour, 64000 Pau, France

Abstract. Full waveform inversion (FWI) of seismic or Ground Penetrating Radar data provides high-resolution quantitative images of the constitutive parameters of the rock/soil which control seismic/GPR wave propagation. We developed a 2D inversion tool in the frequency domain adapted to the multi-parameter physics controlling GPR propagation in isotropic non dispersive media, i.e. dielectric permittivity and electrical conductivity. This inversion engine was previously tested using synthetic 2D data to mitigate the trade-off between the two parameter classes. In this paper, we present the required processing techniques and first inversion results obtained on a real GPR dataset acquired in carbonates with a cross-hole configuration. The presence of the 2 m diameter underground gallery at depth constitutes a nice target to test the robustness, efficiency and resolution of the inversion in such high-contrasts context. Starting from a time tomographic image for the dielectric permittivity and from a homogeneous conductivity, we show that FWI is efficient to retrieve high resolution images of dielectric permittivity but struggles with electrical conductivity. As a quality control, we compare real and synthetic radargrams computed from the tomography and final images, showing the efficiency of the process to reconstruct some events but also underlying some issues, particularly on large incidence angles amplitude traces.

1. Introduction

The knowledge of the subsurface is crucial for a wide range of scientific issues such as environmental, archaeological, hydrological, geological and civil engineering. In this frame, Ground Penetrating Radar (GPR) plays a key role thanks to its non-invasive and high resolution features. Multioffset measurements may provide quantitative information of the subsurface through velocity analysis or inverse processes like traveltime tomography or full waveform inversion (FWI). With traveltime tomography, only a small fraction of the recorded data is involved in the inversion providing a smooth quantitative image of the dielectric permittivity with a limited resolution. In this context, FWI, which takes benefit from the whole recorded signal, should improve the image resolution.

FWI of cross-hole GPR datasets has shown its efficiency on synthetic and real [1] cases. Here, we perform FWI in carbonate environment with the favorable cross-hole configuration in order to evaluate our multi-parameter inversion algorithm. Based on previous

[a] Corresponding author: hugo.pinard@univ-grenoble-alpes.fr

synthetic tests, we use the l-BFGS algorithm which is well known in seismic to reconstruct simultaneously subsurface parameters.

Here, we briefly explain the concepts underlying the FWI with the description of the forward and inverse problems. We then describe the pre-processing methodologies used to estimate the source wavelet and to convert real data acquired in a 3D environment to 2D data managed by the code. Finally, we present our preliminary FWI result which contains many artifacts. Despite these artifacts, which emphasize the current failure of the inversion, we show the first benefits of the method.

2. Theoretical basics on forward and inverse problems

In 2D, the introduction of the constitutive properties in the Maxwell-Faraday and Maxwell-Ampère equations gives two distinct systems: the transverse electric (TE) and the transverse magnetic (TM) modes. Here, we consider only the TM one with vertical electric sources, as used in boreholes. The system of three first-order equations is recast into one second-order equation and two first-order equations. This transformation and acoustic/electromagnetic analogy [2] allow to simulate electromagnetic wave propagation using finite-difference frequency-domain schemes initially developed for seismic modeling in visco-acoustic approximation [3]. This formulation leads to the system $A(m)H_y = s$ where the discrete wave operator $A(m)$ contains the subsurface parameters m while H_y and s are respectively the magnetic wavefield and the discretized source terms. This linear system is solved using a LU factorization through a sparse direct solver [4]. The electrical fields E_x and E_z, associated to GPR antennas, are deduced by finite-difference of H_y following the two first-order equations.

This frequency domain formulation allows to solve TM equations at a reasonable computational cost incorporating easily the diffusive attenuation through the complex effective permittivity $\varepsilon_e = \varepsilon + i\sigma/\omega$ with the dielectric permittivity ε [F/m], the electrical conductivity σ [S/m], the angular frequency ω [rad/s] and the pure imaginary number (using a conventional time-harmonic dependency in $e^{-i\omega t}$).

The formalism of FWI has been introduced in the seismic context by [5] and [6] in the time domain and developed later by [7] in the frequency domain. The target of the FWI is to find the distribution of parameters m which minimizes the real misfit function $f(m)$. This quantity is defined as the distance between the observed data $d_{obs,\omega_j,s}$ and the calculated data $d_{cal,\omega_j,s}(m)$, namely the residuals $\Delta d_{\omega_j,s}(m) = d_{obs,\omega_j,s} - d_{cal,\omega_j,s}(m)$, through the ℓ_2 norm:

$$f(m) = \frac{1}{2}\sum_{j=1}^{N_\omega}\sum_{s=1}^{N_s} \Delta d_{\omega_j,s}(m)^{\dagger}\Delta d_{\omega_j,s}(m) \quad (1)$$

where the symbol † denotes the transpose conjugate operator while the total number of sources and discrete frequencies are denoted by N_s and N_ω respectively. In our particular case, observed and calculated data correspond to the vertical components of the electric field (Ez) at receiver positions. The FWI procedure iteratively updates the model of the subsurface through the formula:

$$m_{k+1} = m_k + \alpha_k \Delta m_k, \quad (2)$$

where α_k is the linesearch parameter satisfying the Wolfe conditions and Δm_k is the descent direction solution of the Newton equation $H_k\Delta m_k = -\nabla f_k$. Among the different methods to solve the Newton equation, we choose the l-BFGS one for its efficiency and accuracy. With this method, the inverse of the Hessian operator $H(m_k)^{-1}$ is approximated through finite-difference of l previous gradients $\nabla f(m_{k-1})...\nabla f(m_{k-l})$. Thus, the algorithm only requires to compute the gradients of each parameter. These quantities are obtained very efficiently thanks to the first-order adjoint-state method ([8] for a review). In the following application, we also introduce the Tikhonov regularization to smooth as much as possible the model update.

3. Application

Data were acquired in an Urgonian karstic limestone massif located near the LSBB. This underground structure goes through the mountain (Fig. 1a). One of its gallery, located close to the exit (S), is relatively shallow and can thus be investigated from boreholes. Among the different boreholes drilled to study carbonate properties, two of them surrounded the gallery (Fig. 1b). This configuration allowed to perform measurements in cross-hole configuration.

Two 100 MHz antennas have been used to record the dataset between these two 50 m deep boreholes separated by 18 m. 44 sources are evenly distributed (1 m interval) in borehole F5 while 200 traces have been recorded every 25 cm in borehole F6. Examples of radargram are displayed in Fig. 1c and d for two different source positions. The first one shows diffractions only between traces 70 and 100 while the second one is more complex, exhibiting three distinct hyperbolae with apex at traces 120, 135 and 145.

All first arrivals have been picked and inverted using a SIRT algorithm to get a smooth image of the velocity and so of the dielectric permittivity (Fig. 2a). The 2 m diameter underground gallery at depth 34 m is visible but appears quite spread compared to the reality where sharp contrasts exist at the interfaces air/concrete/carbonates. We expect that FWI provides higher resolution images in particular of the gallery.

Before applying FWI, a pre-processing of the data is required. We first remove the average value of each trace and apply a Butterworth filter to cut low and high frequency noises. Since our forward modeling engine is 2D and the observed data ($d_{obs}^{3D}(t)$) are 3D, we apply a 3D-to-2D conversion to get the 2D observed data ($d_{obs}^{2D}(t)$). The conversion, given here for one trace, is performed through the formula,

$$d_{obs}^{2D}(t) = \sqrt{\frac{2\pi v(t)^2 t}{\omega}} e^{\imath\pi/4}\, d_{obs}^{3D}(t) \tag{3}$$

with t denotes the time, w the angular frequency, i imaginary number and $v(t)$ the wave velocity which is assumed to be homogeneous (in the formula) and equals to 8 cm/ns following the traveltime tomography result. One example of this conversion is displayed in Fig. 1e where changes in phase and amplitude can be clearly seen. Then, the dataset is normalized by shot-gather to be sure that none of these shot-gathers dominates the misfit function. Note that frequency content of the signal is extracted and displayed in Fig. 1f. In general, all frequencies between 50 to 150 MHz have a strong amplitude so that this frequency band can be used to perform the FWI.

The last step of the pre-processing consists in estimating the source wavelet. In the frequency-domain, this estimation, detailed here for one frequency ω_j and one source s, is classically realized through the linear inversion proposed by [9],

$$S_{\omega_j,s} = \frac{g^{\dagger}_{\omega_j,s}\, d^{2D}_{obs,\omega_j,s}}{g^{\dagger}_{\omega_j,s}\, g_{\omega_j,s}}, \tag{4}$$

where $S_{\omega_j,s}$ is the complex value of the source wavelet at frequency ω_j while $g_{\omega_j,s}$ and $d^{2D}_{obs,\omega_j,s}$ denote respectively the incident Green's functions recorded at receiver positions and the observed data in the frequency domain. Here, each source s is supposed independent from each other so that one source estimation is done per shot-gather. After computations of $S_{\omega_j,s}$ for all frequencies and inverse Fourier transform, the resulting source wavelet for one

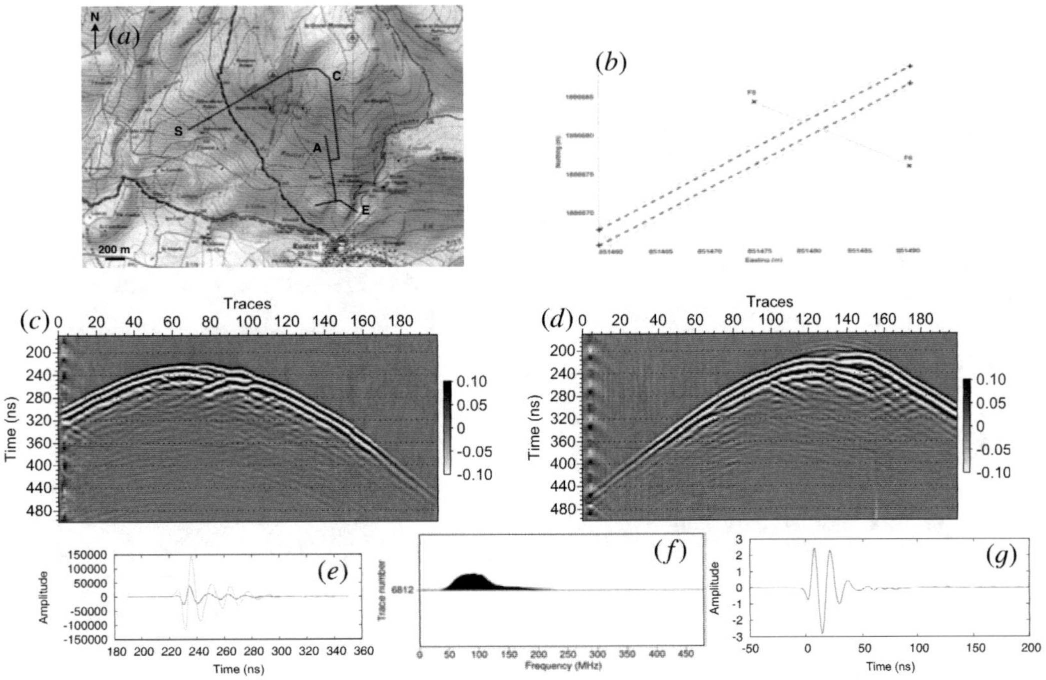

Figure 1. (a) Location of LSBB galleries (black line). E denotes the main entrance, A the anti blast gallery, C the shielded LSBB capsule and S the exit. (b) Zoom close to the exit with the locations of the two boreholes of interest and the gallery. Observed radargrams for source positions at depth (c) 18 m and (d) 34 m. (e) 3D-to-2D conversion for one trace the 3D (red) and 2D (green) traces. (f) Frequency content of one trace. (g) Source wavelet estimated before the inversion for one shot-gather.

source located at depth 17 m is displayed in Fig. 1g. The wavelet may be compared with a Ricker wavelet (second-order derivative of a Gaussian distribution) even if side rebounds are currently quite strong. Even if the wavelet can be updated regularly during the inversion, we choose to estimate it once at the beginning and keep fixed after.

The 51.5 m by 18.5 m model is discretized using a grid interval of 0.05 m to ensure that electromagnetic modeling satisfies the sampling criterion of four/five grid points per minimum wavelength. The initial model for the permittivity is provided by the traveltime tomography while the conductivity one is taken homogeneous (Fig. 2a and d).

The discrete frequencies strategy used to invert data follows the strategy developed by [10] and the frequency content of the dataset (Fig. 1f). One example of this strategy is presented in Table 1. Frequencies of a given group are used to perform one nonlinear inversion and associated results are used as initial models for the next group. Starting by the lower frequencies to reconstruct the large structures of the medium, higher frequencies are progressively introduced to improve the resolution of the model. Note that the frequency sampling ensures a continuous sampling in the wavenumber space. In this paper, we restrict results to the first group because we need to improve some points before using the high frequency content.

As discussed by [11], we invert relative parameters instead of parameter themselves so that they are defined as $\varepsilon_r = \varepsilon/\varepsilon_0$ where $\varepsilon_0 = 8.85 \, . \, 10^{-12}\ Fm^{-1}$ (vacuum value) and $\sigma_r = \sigma/\sigma_0$ where σ_0 has to be chosen accurately. If this parameter is too high, only the

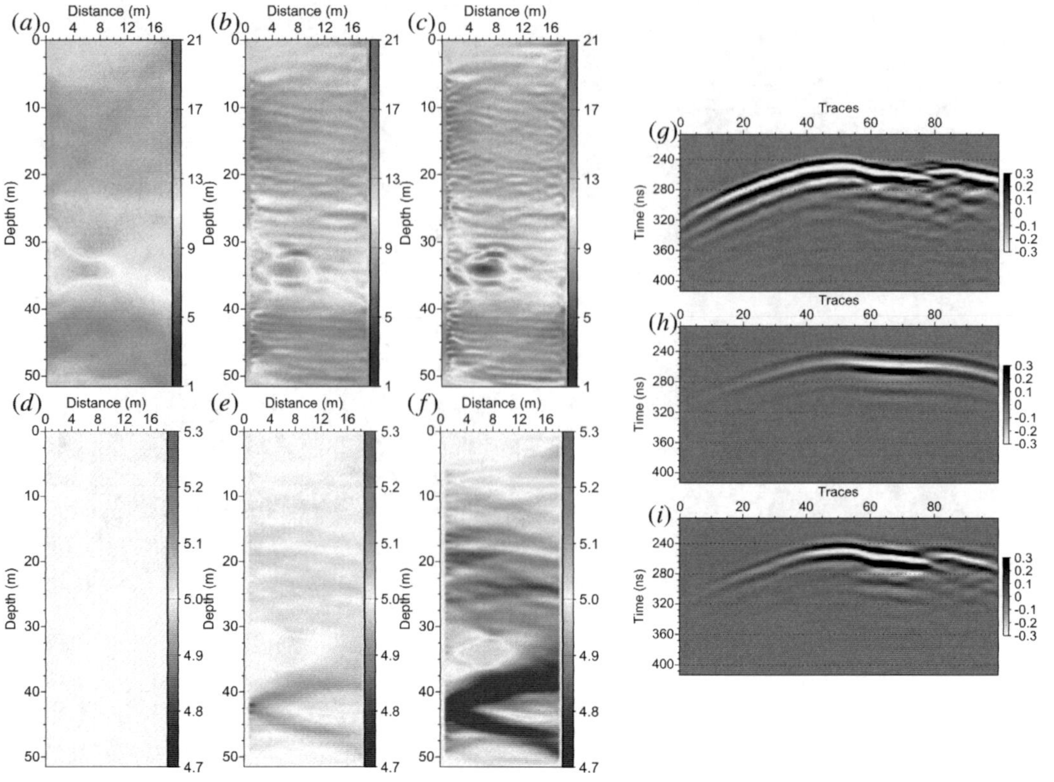

Figure 2. (a–c) Relative permittivity after 0, 15 and 30 iterations; (d–f) conductivity [mS/m] after 0, 15 and 30 iterations. (g) observed data, (h) initial data and (i) final data for 100 receivers associated to one source located at depth 39 m.

Table 1. Frequency strategy for the inversion (MHz).

Group 1	50	60	70
Group 2	60	70	80
Group 3	70	80	100
Group 4	80	100	120
Group 5	100	120	150

conductivity is updated and vice versa for the permittivity. In this paper, $\sigma_0 = 1.0\ 10^{-4}\ S/m$ to slightly promote the relative permittivity update which stabilizes the inversion. Areas around sources and receivers have been frozen to prevent the model being updated only at these positions. We decided to perform 30 nonlinear iterations. The normalized misfit functions with respect to the initial value are 0.8 and 0.78 respectively at the 15th and 30th iterations. This small decrease is linked to difficulties for the algorithm to remove the imprint of sources and receivers despite the presence of the frozen zone (Fig. 2). At the end, sources and receivers, through their vertical dipolar injection points, create horizontal artifacts in the model. These latter, which present a higher wavenumber content in the permittivity model, still require a deeper analysis to better understand their origins and remove them.

Despite these artifacts, we obtain an increase of the resolution of the gallery focusing the permittivity and providing a first image of conductivity. In Fig. 2, the limit of the gallery becomes progressively sharper while the final relative permittivity value in the gallery is

DOI: 10.1051/e3sconf/20161201002

closer to the expected one ("1" for the void). For the conductivity model, the gallery appears more conductive than the geological medium and the presence of water-saturated layers around 38 m depths is not detected. These results can be explained either by the presence of iron in the concrete of the gallery (even if its presence is not confirmed) or more probably by an initial model of conductivity too simple and/or too far from the true one.

Because the FWI is based-on data fitting procedure, radargrams are shown in Fig. 2g, h and i. Comparing observed and initial data, several differences can be underlined. Before trace 45, a clear mismatch in amplitude is visible while after this trace, the initial data do not reproduce the complex structure of the observed data and specially the discontinuity around trace 77. The analysis of the final radargram shows that the mismatch in amplitude is still present meaning that the algorithm did not update the model to reduce this difference. Actually, the algorithm concentrates its efforts on the complex arrivals and the discontinuity which are well retrieved at the end of the inversion. This final radargram, obtained with only one group of frequencies, emphasizes the potential of the FWI.

4. Conclusion

We briefly propose the formulation of a 2D frequency-domain FWI for GPR datasets in TM configuration to reconstruct simultaneously dielectric permittivity and electrical conductivity. Then, we present our first results using a dataset recorded near the LSBB. These results are promising with a focalization of the gallery in permittivity and a first image of conductivity. This work will be continued with a better handling of the source estimation (at each nonlinear iteration), the consideration of all groups of frequencies and other initial conductivity models. These different points should improve the quality and the resolution of the final reconstructions and, we hope, allow the detection of multi-scale heterogeneities characterizing the carbonate properties of this massif. Final results will be compared to well logs, electrical tomography and checked against ground truth (for the gallery location). At middle/long term, new acquisitions will be performed and results will be used to follow the evolution of the massif properties over time.

This study was partially funded by the SEISCOPE consortium (`http://seiscope2.osug.fr`), sponsored by BP, CGG, CHEVRON, EXXON-MOBIL, JGI, PETROBRAS, SAUDI ARAMCO, SCHLUMBERGER, SHELL, SINOPEC, STATOIL, TOTAL and WOODSIDE. This study was granted access to the HPC resources of CIMENT infrastructure (`https://ciment.ujf-grenoble.fr`) and CINES/IDRIS under the allocation 046091 made by GENCI. We thank the LSBB facilities for their support for field experiments.

References

[1] A. Klotzsche, J. van der Kruk, G.A. Meles, H. Vereecken, Geophysics, **77**, H57 (2012).

[2] J.M. Carcione and F. Cavallini, Wave motion, **21**, 149 (1995).

[3] B. Hustedt, S. Operto, and J. Virieux, Geophysical Journal International, **157**, 1269 (2004).

[4] MUMPS-team. *MUMPS – Multifrontal Massively Parallel Solver users' guide – version 4.9.2 (November 5, 2009)*. ENSEEIHT-ENS Lyon, `http://www.enseeiht.fr/apo/MUMPS/` or `http://graal.ens-lyon.fr/MUMPS` February (2009).

[5] P. Lailly, *The seismic problem as a sequence of before-stack migrations*, in *Conference on Inverse Scattering: Theory and Applications*, edited by J. Bednar, (SIAM, Philadelphia, 1983).

[6] A. Tarantola, Geophysics, **49**, 1259 (1984).
[7] R.G. Pratt and M.H. Worthington, Geophysical Prospecting, **38**, 287 (1990).
[8] R.E. Plessix, Geophysical Journal International, **167**, 495 (2006).
[9] R.G. Pratt, Geophysics, **64**, 888 (1999).
[10] L. Sirgue, R.G. Pratt, Geophysics, **69**, 231 (2004).
[11] F. Lavoué, R. Brossier, L. Métivier, S. Garambois, J. Virieux, Geophysical Journal International, **197**, 248 (2014).

Atmospheric radiation environment analyses based-on CCD camera at various mountain altitudes and underground sites

Pierre Li Cavoli[1,2,a], Guillaume Hubert[1,b], and José Busto[2,c]

[1] The French Aerospace Laboratory (ONERA), 31000 Toulouse, France
[2] The Center for Particle Physics of Marseilles (CPPM), CNRS/IN2P3, 13288 Marseilles, France

Abstract. The purpose of this paper is to discriminate secondary atmospheric particles and identify muons by measuring the natural radiative environment in atmospheric and underground locations. A CCD camera has been used as a cosmic ray sensor. The Low Noise Underground Laboratory of Rustrel (LSBB, France) gives the access to a unique low-noise scientific environment deep enough to ensure the screening from the neutron and proton radiative components. Analyses of the charge levels in pixels of the CCD camera induced by radiation events and cartographies of the charge events versus the hit pixel are proposed.

1. Background and context

The Earth's atmosphere is constantly bombarded by cosmic rays which come from the Sun and the space. These cosmic rays are essentially relativistic protons (85%), helium nuclei (13%) and heavy nuclei (2%) [1]. When the cosmic rays penetrate the atmosphere, they interact with its constituents such as nitrogen and oxygen atoms. These interactions generate cascades of secondary radiation of every kind called secondary particles [2]. Protons, neutrons, electrons, positrons, muons, pions and heavy nuclei are the main products [3]. Their intensity depends on the altitude, the geomagnetic latitude and the Sun's activity. Among the secondary particles which reach the earth surface, muons are the most abundant [4]. Historically the ionizing power of muons was not took into account because of their weak interaction with matter. Nevertheless with the current nanometrical scales of electronics devices, the muons ionization wake can no longer be neglected [5]. Radiations induced Single Event Effect (SEE) have been identified as a major reliability issue for commercial electronic systems [3, 6, 7]. These perturbations are the consequence of a current pulse generated by the impact of energetic particles in sensitive electrodes of the circuit. Soft errors are caused by cosmic rays and alpha particles induced by the natural alpha emitters from materials. As process node scales down, soft errors affect increasingly electronic systems including circuits from advanced CMOS technologies. Soft errors, which appear at the end of the 60's for space applications, constitute nowadays a major challenge not only for embedded systems of space, and avionic applications, but also for any system operating at ground level [8] for

[a] e-mail: pierre.li_cavoli@onera.fr
[b] e-mail: guillaume.hubert@onera.fr
[c] e-mail: busto@cppm.in2p3.fr

which a fault may have critical consequences. The constant progresses in integrated circuits manufacturing technologies result in significant reductions of transistor dimensions. This makes very large-scale integration circuits more and more sensitive to perturbations. Thus a set of new error mechanisms appear. With such advanced technologies the track effect, namely the radial ionization profile, is expected to be critical multiple-cell upsets being the main contributor [9–11]. As feature sizes scales down, charge collection at multiple nodes (i.e., charge sharing) due to a single particle hit results from the decrease spacing dimensions between cells [12–15]. Then the sensitivity to SEE is expected to increase and recent studies have shown the occurrence of soft-errors due to protons [16, 17] and to muons [18] for advanced nano-scales electronics. In order to know the muon contribution in soft-errors, the first requirement is to ensure a solid understanding of the muon radiation environment. That is why through this paper we will present the feasibility of characterizing the atmospheric muon environment with a CCD camera. Then we will describe the analyses of the CCD camera measurements run in atmospheric sites and also in low noise environment in the underground facilities provided by the LSBB and the Modane Underground Laboratory (MUL).

2. Camera characterization

2.1 Low-noise characterization

The PIXIS 2048 CCD camera from the Princeton Instruments series is a 4.2 million-pixel (2048 × 2048) of 13.5 μm^2 full frame CCD with an imaging area of 7.64 cm^2 and a depletion region of 4 μm. The imaging process of the observation of an object is based-on the shifting of the charge collected in each pixel of the CCD matrix. When a charged particle crosses the imaging area, a charge level is induced in each pixel which was hit along the particle track. The total charge deposit by the incident particle can be restored by summation of all fractional charges in each adjacent hit pixel [19]. A first measurement has been made in the Modane Underground Laboratory in order to characterize the intrinsic camera noise. Burried at a depth of 4800 m.w.e, only very few secondary particles from cosmic rays succeed in reaching the laboratory. A lead castle has been placed all around the camera for shielding it from remaning radioactivity and rare secondary particles. The integrated fluence ϕ_{int} is calculated as the sum of the detected particles divided by the CCD sensitive surface S_{pixel} for four days of measurement (Eq. 1) and is displayed as a function of the exposition number (Fig. 1).

$$\int_1^{N_{expo}} \phi_{int} = \frac{\sum_{i=1}^{N_{expo}} detected\ particles\ (i)}{S_{pixel}} \tag{1}$$

with 2821 detected particles by cm^2 for four days of measurement and $S_{pixel} = 7.64\,cm^2$. Each exposition records charge pixels during 50 seconds before be cleaning out. From the integrated fluence the mean flux is extracted and give access to the intrinsic noise as (Eq. 2):

$$\bar{\Phi}_{flux_{CCD}} = \frac{\int_1^{N_{expo}} \phi_{int} \cdot T_{expo}}{N_{expo}} \tag{2}$$

with $\int_1^{N_{expo}} \phi_{int}$ = 369 particles.cm^{-2}, T_{50s} = 72 (number of expositions for one hour of measurement), N_{expo} = 6912 expositions (for four days of measurements). This intrinsic noise is equal to 3.8 particles.cm^{-2}.h^{-1} and is substracted for the next measurements.

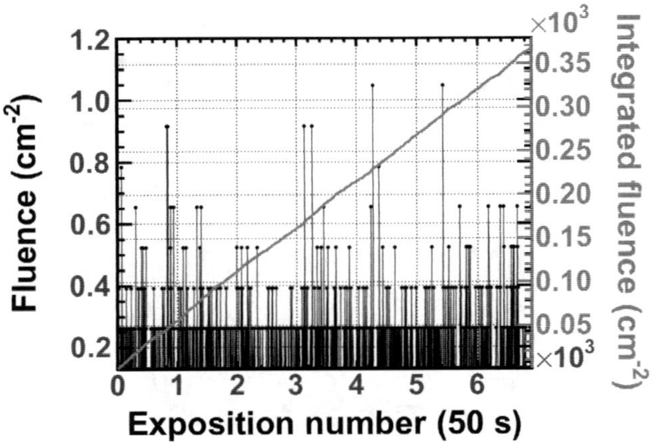

Figure 1. Fluence versus exposition number at the Modane Underground Laboratory.

2.2 Radon and gamma irradiations

In an underground location, particles emitted due to natural radiation can interact with the detector. Because the CCD camera detect all kind of ionizing particles and is not directly able to ascertain their nature, we have investigated the camera sensitivity to the products of the natural radioactivity as the radon and gamma. In laboratory, we have assembled a cuve fill of radon by an injecting pump with the CCD camera placed inside in a closed-circuit system. The time of the experiment has been limited to two hours in order to prevent the overheating of the camera. The detected particles flux is displayed with and without radon (Fig. 2a). A mean flux of 70 particles.cm^{-2}.h^{-1} is obtained with radon namely 25 particles.cm^{-2}.h^{-1} more than with the empty cuve. Thus the CCD camera is sensitive to the impact of radon. In the same manner, the camera has been placed through contact with a light radium source (^{226}Ra) which emit gamma, alpha and beta through the ^{238}U decay chain. The CCD camera is also sensitive to gamma rays by detecting 47 particles.cm^{-2}.h^{-1} more than with the reference measurement (Fig. 2b).

3. Atmospheric and underground measurements

3.1 Detected particle flux analyses

After the camera characterization we used the LSBB for atmospheric and underground sites. For the first kind of measurements, we had access to "Vestale" which is a mountain location at the summit of the "Grande Montagne" perched at 1067 m above the sea-level [20]. The detected particles flux obtained at "Vestale" is compared with the one obtained at the Center for Particle Physics of Marseilles (CPPM). Their mean flux is 161 particles.cm^{-2}.h^{-1} and 90 particles.cm^{-2}.h^{-1} respectively for an altitude of +1067 and +150 m (Fig. 3). For the underground measurements, the LSBB has allowed us to use two rooms of different depth. The first one is buried at 60 m under the rocks at the laboratory entrance. The second one, which is called "Rameau" is the deepest part of the LSBB and is buried at a depth of 500 m. A first measurement was made in 2011 (Fig. 4). The detected particle flux at 500 m is greater than the flux at 60 m. This result can be explained by the presence of radon during the measurement. Indeed the radon radioactivity increases with the depth and with the difference between the temperature in the buried room and the temperature at the ground surface. When the temperature on the top of the mountain is greater than in the galleries of the LSBB, the

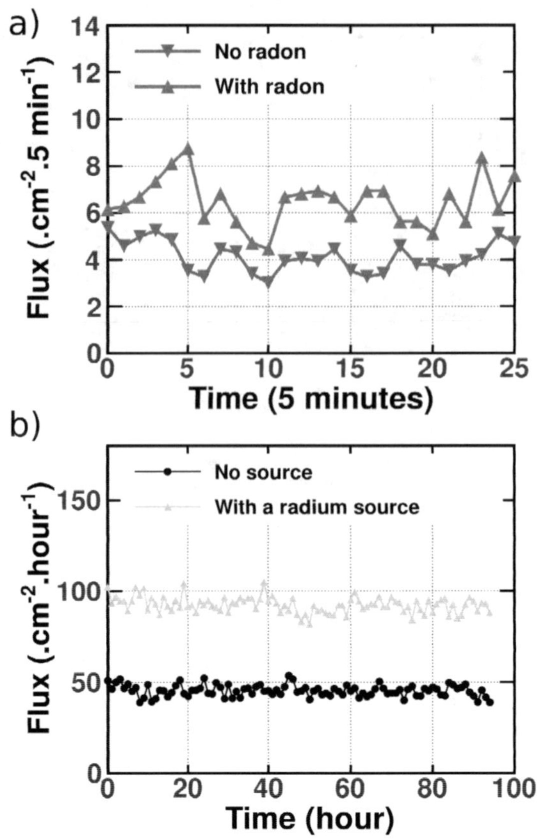

Figure 2. Detected particle flux: (a) in the presence of radon (b) with a radium source.

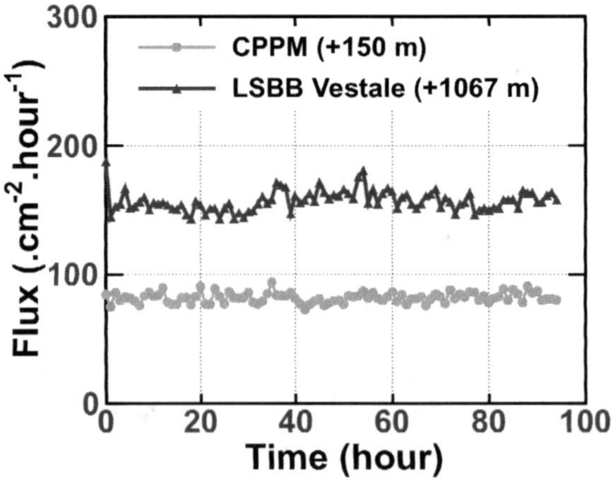

Figure 3. Detected particle flux for the atmospheric sites.

radon tends to be released from the rocks and go down toward the galleries. This situation occurs when the outside temperature reaches nearly 15°C. Nevertheless, in 2012 a ventilation system was set up in order to reduce the radon flux. This system was efficient and allowed to decrease the flux of the detected particles by a factor of 2.4, as it is visible from the new measurement realized in 2015 (Fig. 4).

DOI: 10.1051/e3sconf/20161203003

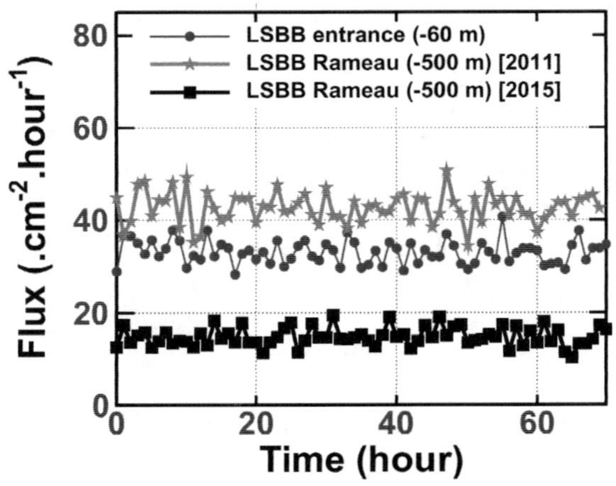

Figure 4. Comparison of the detected particle flux with (2015) and without (2011) ventilation system.

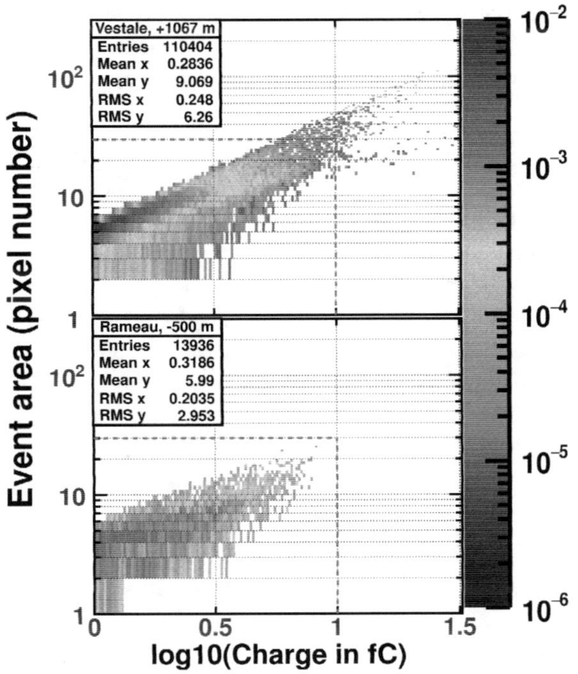

Figure 5. Event cartography: (a) atmospheric site (+1067 m) (b) underground site (−500 m).

3.2 Event cartography analyses

From these measurements the event cartography of "Vestale" and "Rameau" (respectively +1067 m and −500 m) are compared (Fig. 5). The scale colors represents the occurence probability of the event. Both cartographies are normalized on four days. At +1067 m, 110 404 events are detected whereas at −500 m only 13 936 events are. The red dashed lines (at 10 fC and 30 hit pixels) represent the limited area detection by the underground measurement site: outside this area no events are detected at −500 m. This cut corresponds to the absorption of the hadronic component (mainly protons and neutrons) by the depth of rocks. Because muons are the only particles which can penetrate such a depth (except the neutrino that CCD camera can not detect), only muons and the irreductible natural radioactivity which can not be vanished remains at −500 m. The red core visible at +1067 m located in a zone which spread from 4 to 10 hit pixel and from 1 to 3 fC, represents the most probable events and is also visible at −500 m but it is more diffused. Because muons are the most abundant secondary particles at the ground surface [4], this core can be attributed to the muon charge deposit pattern.

4. Conclusion

Through this paper we have described charge level analyses from a CCD camera utilized as a cosmic ray sensor. Thanks to several different depths of rocks and a high-mountain altitude site, the LSBB appeared to provide a remarkable measurement site for investigating secondary particles from cosmic rays. Combined with measurements at the MUL, the analyses of the events observed by the CCD camera give access to the detected particles flux and the charge deposition pattern. Results showed that high charge level events seen on atmospheric sites can be considered as hadronic component (mainly neutrons and protons) while low charge levels and punctual events are induced by muons which are able to generate up to 3 fC in the CCD camera. Hence, thanks to double level of measurement sites, muon discrimination from other secondary particles has been investigated.

References

[1] G. Reitz, Radiation Protection Dosimetry (1993).
[2] J.L. Barth, C. Dyer, E. Stassinopoulos, Nuclear Science, IEEE Transactions on **50**, 466 (2003).
[3] J.F. Ziegler, IBM journal of research and development **40**, 19 (1996).
[4] K. Nagamine, *Introductory muon science* (Cambridge University Press, 2003).
[5] B.D. Sierawski, Mendenhall et al., Nuclear Science, IEEE Transactions on **57**, 3273 (2010).
[6] D. Lambert, J. Baggio, G. Hubert, V. Ferlet-Cavrois, O. Flament, F. Saigne, F. Wrobel, H. Duarte, J. Boch, B. Sagnes et al., Nuclear Science, IEEE Transactions on **52**, 2332 (2005).
[7] D. Lambert, J. Baggio, G. Hubert, P. Paillet, S. Girard, V. Ferlet-Cavrois, O. Flament, F. Saigne, J. Boch, B. Sagnes et al., Nuclear Science, IEEE Transactions on **53**, 1890 (2006).
[8] G. Hubert, R. Velazco, C. Federico, A. Cheminet, C. Silva-Cardenas, L. Caldas, F. Pancher, V. Lacoste, F. Palumbo, W. Mansour et al., Nuclear Science, IEEE Transactions on **60**, 2418 (2013).
[9] A. Akkerman, J. Barak, D. Emfietzoglou, Nuclear Instruments and Methods in Physics Research Section B: Beam Interactions with Materials and Atoms **227**, 319 (2005).
[10] R.K. Lawrence, A.T. Kelly, Nuclear Science, IEEE Transactions on **55**, 3367 (2008).
[11] G. Hubert, P. Li Cavoli, C. Federico, L. Artola, J. Busto, IEEE Transactions on Nuclear Science **62**, 2837 (2015).
[12] O.A. Amusan, A.F. Witulski, L.W. Massengill, B.L. Bhuva, P.R. Fleming, M.L. Alles, A.L. Sternberg, J.D. Black, R.D. Schrimpf, Nuclear Science, IEEE Transactions on **53**, 3253 (2006).
[13] B.D. Olson, O.A. Amusan, S. Dasgupta, L.W. Massengill, A.F. Witulski, B.L. Bhuva, M.L. Alles, K.M. Warren, D.R. Ball, IEEE Transactions on Nuclear Science **4**, 894 (2007).

[14] A.M. Francis, D. Dimitrov, J. Kauppila, A. Sternberg, M. Alles, J. Holmes, H.A. Mantooth, Nuclear Science, IEEE Transactions on **56**, 3109 (2009).
[15] J. Black, D. Ball, W. Robinson, D. Fleetwood, R. Schrimpf, R. Reed, D. Black, K. Warren, A. Tipton, P. Dodd et al. (2008).
[16] B.D. Sierawski, J.A. Pellish, R.A. Reed, R.D. Schrimpf, K.M. Warren, R.A. Weller, M.H. Mendenhall, J.D. Black, A.D. Tipton, M.A. Xapsos et al., Nuclear Science, IEEE Transactions on **56**, 3085 (2009).
[17] G. Hubert, S. Duzellier, F. Bezerra, R. Ecoffet, in *Radiation and Its Effects on Components and Systems (RADECS), 2009 European Conference on* (IEEE, 2009), pp. 179–186.
[18] G. Hubert, L. Artola, D. Regis, Integration, the VLSI journal **50**, 39 (2015).
[19] G. Hubert, A. Cheminet, T. Nuns, V. Lacoste, Nuclear Science, IEEE Transactions on (2013).
[20] G. Waysand, et al., NIMPR Section A (2000).

E3S Web of Conferences **12**, 04004 (2016)
DOI: 10.1051/e3sconf/20161204004
i-DUST 2016

Toward an innovative stochastic modeling of electric charges loss through dielectric

G. Micolau[1], J. Postel-Pellerin[2], P. Chiquet[2], M. Joelson[1], C. Abbas[1], D. Boyer[3], and C. Ginoux[1]

[1] UMR 1114 EMMAH, University of Avignon, Department of Physics, Campus Agroparc, 301 rue Baruch de Spinoza, BP. 21239, 84916 Avignon Cedex 9, France
[2] UMR 7334 IM2NP, University of Marseille, 5 rue Enrico Fermi ou 60 rue F. Joliot Curie, Technopôle de Château Gombert, 13453 Marseille Cedex 13, France
[3] UMS 3538 LSBB, La Grande Combe, 84400 Rustrel, France

Abstract. This paper deals with new stochastic modeling of very low tunneling currents in Non-Volatile Memories. For this purpose, we first develop current measurement method based on Floating Gate technique. In order to reach the long time behavior of electrical dynamic, we aim at using very basic tools (power supply, multimeter...) but still having a very good current resolution. Also, our measurement is led in a very particular low-noise environment (underground laboratory) allowing to keep the electrical contacts on the device under test as long as possible. After showing the feasibility of such measurements, we present a modeling approach of the charge loss process inside the Non-volatile Memories by using mathematical tool involving long memory effect. The model is based on stochastic counting process with memory effect yielding to a fractional relaxation equation for the charge loss over time. The main interest of the present model lies in the fact that the corresponding inversion problem involves only two parameters that can be carried out efficiently.

1. Introduction

Flash memory cells are based on the floating gate technology principle [1]. The most widespread solution to enable semiconductor memories to be non-volatile, that is to say able to keep information without any power supply, is to use MOS transistors whose threshold voltage is shifted by a charge stored in an isolated gate above the channel. Floating gate technologies consist in adding a second gate between the gate and the channel of a classical MOS transistor. This second gate can isolate charges to make the transistor threshold voltage variable. Most of the time, charges are injected through a dielectric, in general Silicon dioxide SiO_2, placed between the floating gate and the transistor channel, as presented in Fig. 1. Lastly, the two gates of this "transistor" are separated by another dielectric, most commonly a tri-layer stack oxide "Oxide/Nitride/Oxide" (ONO). Thus, the Flash elementary cell, constituted by a floating gate transistor called "state transistor", can be seen as a classical MOS transistor whose gate would be in series with a capacitor C_{pp}. This gate, called "Floating Gate" (FG), can now store a charge while the second electrode of the capacitor becomes the "Control Gate" (CG) of the cell.

Barrier transparency in the tunnel oxide, which can be electrically modeled by a current source I, allows the injection of charges in the floating gate, shifting the MOS transistor

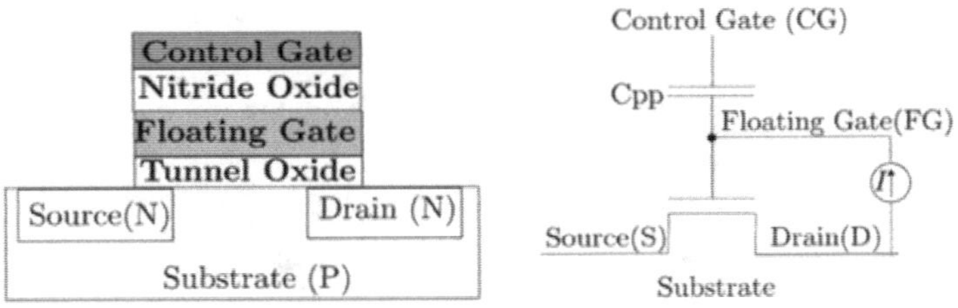

Figure 1. Schematic of a floating gate structure (left) and its electrical scheme (right).

threshold voltage V_T according to (1):

$$\mathbf{V_T = V_{T0} - \frac{Q_{FG}}{C_{pp}}} \tag{1}$$

where V_{T0} is the natural threshold voltage of the cell, Q_{FG} the charge amount in the floating gate and C_{pp} the capacitance between the control gate and the floating gate.

A quantity of charge Q_{FG} is injected into the floating gate by using an adequate set of biases, depending on technology (Flash or EEPROM) [2–5]. Memory cell reliability, defined as the capability of said device to function over time, is a major issue for manufacturers and can be related to many parameters such as process and using conditions. Indeed, unceasing device scaling, decrease of dielectric thicknesses around the floating gate and high voltages applied on cells are many parameters altering memory reliability [6].

Thus, dielectrics' quality is a major issue due to their antagonistic roles: avoiding electric charge leakage currents during retention phase while being transparent enough during programming steps. A better understanding of these leakage currents is crucial to improve the whole quality of our memory cells, that's why we have to develop powerful methods to reach very low current levels.

The "Floating-Gate Technique" (FGT) is generally used and will be detailed in the next section [7, 8]. Using this method requires i) protecting the wafer from mechanical perturbations (vibrations) and ii) (very) long time acquisition. In a classical laboratory environment, those two constraints are generally not easy to deal with. Indeed, they lead to use, for a very long time, huge, heavy and very expensive probers (mechanically insulated by air shocks), which is often economically not possible. The main idea of our experimental platform consists in developing a "cheap" platform. The mechanical insulation is "naturally" done by the very peculiar environment of the underground LSBB Laboratory located in Apt (South of France), presented in Sect. 3.

The second part of the paper is devoted to the presentation of a fractional relaxation modeling to describe the time evolution of the charge loss through the dielectric. As we will see, the main reason to need to model charge loss dynamic lies into the fact that the direct numeral derivation of this latter yields to an unusable leakage current. In the other hand, classical electrical models namely the Fowler-Nordheim and Poole-Franckel are inoperative to model data at the very low range value of leakage current encountered in this work.

2. Floating Gate Technique (FGT)

Most of the electrical measurements developed to characterize semiconductor devices and especially Non-Volatile Memories are based on current measurements. Indeed, to study a major reliability aspect of these NVM we have to evaluate the very low-level leakage currents responsible for the charge loss during the retention phase. Nevertheless, these currents are

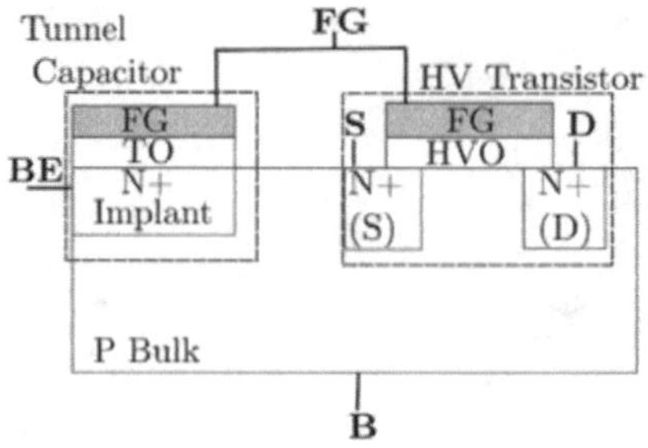

Figure 2. Floating-Gate test structure.

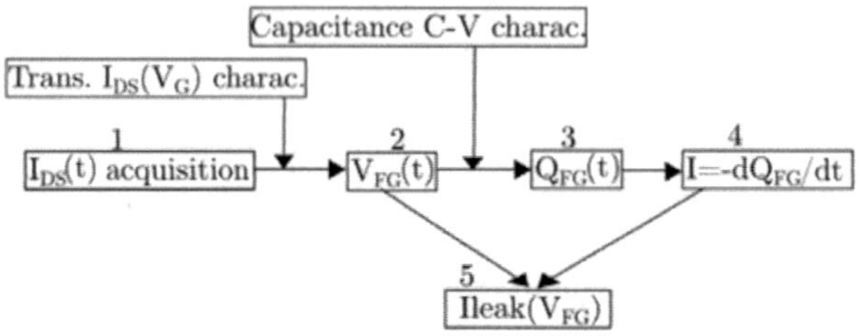

Figure 3. The five key points of Floating-Gate Technique methodology to extract leakage current.

not accessible through direct measurements, even with high-performance analyzers [9], so we have to use some indirect measurement techniques to reach lower level currents. One of the most widely used technique is the "Floating-Gate Technique" (FGT), based on the use of a MOS capacitor and a MOS transistor in parallel [7, 8].

A. Floating-Gate test structure

The Floating-Gate test structure, presented in Fig. 2, consists in a large High-Voltage transistor whose gate is common with the Top Electrode of a large tunnel capacitor, denoted C_{tun}. This common gate plays the role of the Floating-Gate in a memory cell but is directly accessible to apply biases. Tunnel capacitor represents the injection zone of the memory cell. Indeed it has exactly the same process conditions as the injection region (oxide thickness, doping conditions...).

B. Floating-Gate Technique methodology

The Floating-Gate Technique (FGT) is based on the voltage measurement of an initially charged gate of a MOS capacitor, which is then disconnected from the external circuit during the experiment. The measurement of this slowly decreasing gate voltage is performed indirectly through the measurement of the drain current of the transistor sharing its gate with the capacitor. This transistor "converts" the charge of the capacitor, and thus the gate voltage, in a measurable drain current. The temporal variation of this drain current is directly linked to the variation of the gate voltage and thus to its gate charge which is the same as the capacitor charge. Figure 3 depicts the full methodology allowing to obtain the leakage current I_{leak} as a function of the Floating-Gate voltage V_{FG}.

The extraction methodology requires two preliminary characteristics, the $I_{DS}(V_{FG})$ characteristics of the transistor and the C-V curve of the tunnel capacitor, presented in Fig. 4.

 DOI: 10.1051/e3sconf/20161204004

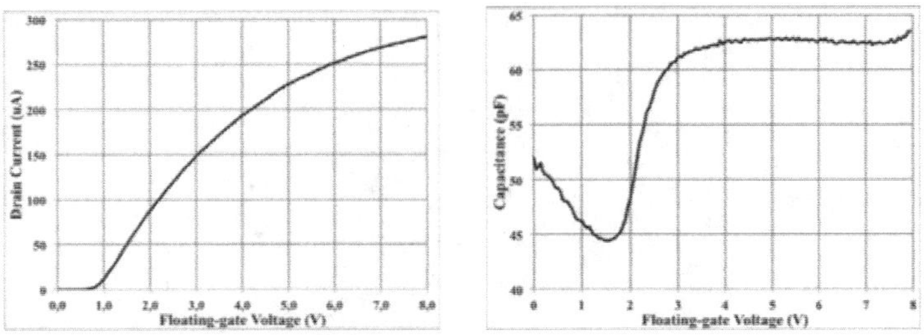

Figure 4. Preliminary characteristics required in the leakage current extraction: $I_{DS}(V_{FG})$ (left) and C_{tun}–V_{FG} (right).

The Floating-gate voltage is the image of the Floating-Gate charge Q_{FG}, extracted from the $C_{tun}(V_{FG})$ characteristics, presented in Fig. 4, using (2):

$$\mathbf{Q_{FG}} = C_{\mathbf{tun}} \times \mathbf{V_{FG}}. \tag{2}$$

The leakage current I_{tun} is then the variation of this Floating-gate charge Q_{FG} over time (3):

$$\mathbf{I_{tun}} = -\frac{\mathbf{dQ_{FG}}}{\mathbf{dt}}. \tag{3}$$

To reach very low-level currents, we have to acquire the drift of the transistor drain current over very long time, keeping applied biases on the transistor (except on the Floating Gate) during the whole experiment. The main difficulty is to keep the electrical contacts on the device under test for days, weeks or even months, knowing the longer the measurement, the lower the extracted current. When using classical probe stations, it is difficult to keep electrical contacts for more than a few days due to the ambient vibrations [10]. Electromagnetic perturbations also disturb the experiment due to the very slow drift of the drain current we have to measure. All the existing solutions to enable a long drain current acquisition lead to very heavy and expensive test bench that we propose to avoid in our study. The improvement consists in developing a cheap but very sensitive test bench, embedded in a particular environment.

3. Low-noise environment

To reach very low levels of electrical and mechanical noises, we have first chosen a specific test environment with a very low electromagnetic noise and a very low vibration level, allowing to use a very simple test bench.

3.1 Specific test environment: The Low Noise Underground Laboratory of Rustrel

The Low Noise Underground Laboratory in Rustrel Pays d'Apt (South of France) is a set of horizontal galleries dug in the bedrock of Big Mountain, bordering South Albion plateau. It was dedicated to be the former cockpit shooting # 1 of the French nuclear deterrent force, from 1973 to 1998. It was built to resist a nuclear weapon assault. It allows access to different rooms (for a total of 14,000 m^2) along 3.7 km, in a rock cover varying from 30 to 519 m under the Surface topography. Rooms and galleries are shielded (electromagnetic waves), by concrete and massive steel shields. This peculiar environment is now used by the National Institute of Universe (INSU) as a hydrogeological, geophysical (net of seismographs) and astronomical (muons detectors) observatory since few years. It also allows to test microelectronic devices in a non radiative environment. For our purpose,

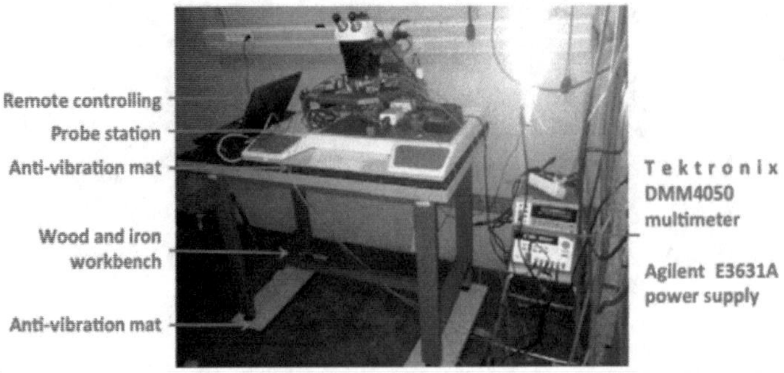

Figure 5. Picture of the proposed test bench, embedded in the low-noise environment.

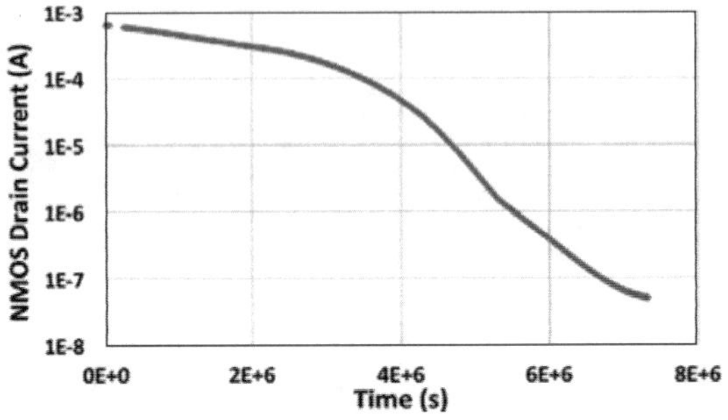

Figure 6. Acquisition of the drain current of the NMOS transistor for an initially positively charged tunnel capacitor.

this laboratory is the "perfect environment" since it allows to avoid electromagnetic and mechanical perturbations during a very long time range. Our platform has been installed at the end of the gallery, where the perturbations are supposed to be the smallest.

3.2 Proposed test bench

Our experimental platform is based on a classical probe station with 5 manipulators and needles, an Agilent E3631A triple output DC power supply and a Tektronix DMM4050 digital multimeter, as illustrated in Fig. 5. To avoid as much as possible any movement around the test bench, a fully remote controlled experiment has been performed. The complete description of the test bench is given in [11] and improvements in the robustness of the technique can be found in [12].

3.3 Instance of experimental results

The experiment exposed here has been successively performed on a NMOS Floating-Gate test structure. The drain current drift acquisition over time (up to almost three months) is shown in Fig. 6.

By applying the methodology detailed in Section II.B and using preliminary characteristics from Fig. 4, we can obtain from the evolution of the capacitor charge Q_{FG} (Fig. 7a) and the extracted leakage current (Fig. 7b) over time.

As a first comment from Fig. 7a, we note that at time around t = 2.5e6 sec. a slope break on the charge behavior appears clearly. This suggests the presence of two schemes of the charge decrease. As observed in Fig. 7b, the curve of the leakage current from direct

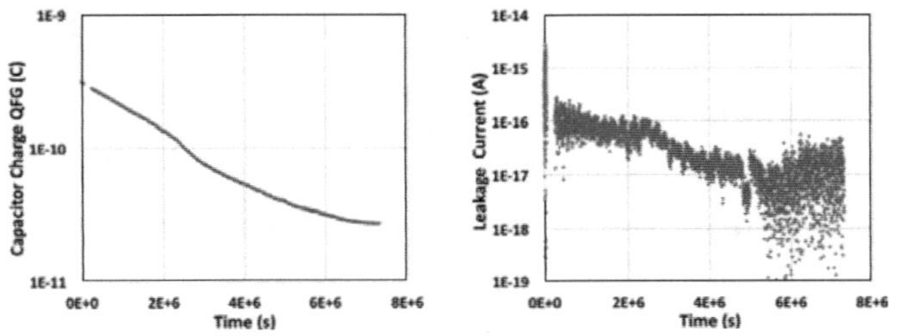

Figure 7. Extracted a) capacitor charge and b) leakage current, on NMOS.

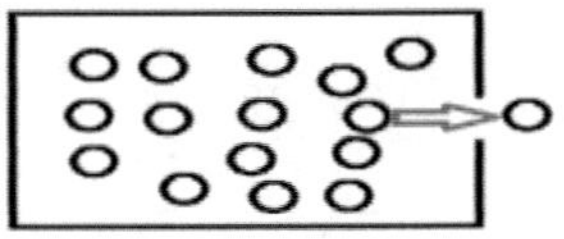

Figure 8. Charges trapped in the dielectric may be seen as a sample of moving particles inside a box with an opening to the outside.

derivative of charge evolution is found to be highly noisy and is not suitable to define a clear trend for current leakage. More precisely, this means that the charge loss curve does not possess the required mathematical derivability properties. This is related to a more deeply reason involving the fact that process causing charge particle to come out the dielectric medium is probabilistic. As a consequence, it is necessary to model the charge evolution curve as a relaxation law of random process.

4. Modeling electric charge dynamic

4.1 Relaxation model

The simplest model of relaxation phenomenon is given by the exponential law. To illustrate it, let us consider that electrons (charges) may be seen as a sample of moving particles inside a box with an opening to the outside.

Then, measuring charge evolution inside the dielectric is equivalent to counting number of remaining particles. Let q(t) the probability that one particle will come out of the box during time interval [0,t[. In addition,one can make the assumption that the output probability q (t) do not change with the number ofremaining particles. In other words, there is no memory effect on the random process. This is the *Poisson process* for which, the probability to have N particles at time t is given by

$$(N,t) = \frac{(kt)^N}{N!}\exp(-\mathrm{Ct}). \tag{4}$$

In this case, it is well-known that the survival probability q(t) writes as:

$$\mathrm{q(t)} = 1 - \exp(-\mathrm{Ct}). \tag{5}$$

The parameter C is the probability that a given particle is instantaneously out side the box. Also, the number of particles inside the box will be a random process N(t) following a binomial law and then the mean value n at time t follows ordinary differential equation

$$\frac{d}{dt}n(t) + Cn(t) = 0. \tag{6}$$

Obviously, the solution is given by an exponential law

$$n(t) = \mathrm{n_0} \exp(-\mathrm{Ct}) \quad (7)$$

where $\mathrm{n_0}$ is the initial number of particles. It is a memory less relaxation law, known also as the desintegration law.

4.2 Relaxation with Memory Effect

If memory effect is taken in account *i.e.* the probability of particle to come out is now depending on the number of remaining particles, then we have a Fractional Poisson Process [16] and the probability P(N,t) is given by

$$\begin{cases} P(N,t) = \frac{(-t)^N}{N!} \frac{d^N}{dz^N} E_{\alpha,1}(\mathrm{z}) \Big| \\ P(N=0,t) = E_{\alpha,1}(-(\mathrm{C}t)^\alpha) \end{cases} \mathrm{z} = -(Ct)^\alpha . \quad (8)$$

Where $E_{\alpha,\beta}(\mathrm{z})$ is the Mittag-Leffler function [18] defined as:

$$E_{\alpha,\beta}(\mathrm{z}) = \sum_{m=0}^{\infty} \frac{z^m}{\Gamma(\alpha m + \beta)} \quad \alpha > 0 \text{ and } \beta \in \mathbf{C} \quad (9)$$

with $\Gamma(\mathrm{t}) = \int_0^\infty x^{(t-1)} e^{-x} dx$ is the Gamma function. The Mittag-Leffler function is a generalization of the exponential function with $E_{1,1}(\mathrm{z}) = \exp(\mathrm{z})$. The real α is the fractional order of the process. The mean value n of number of particles at time t follows the fractional relaxation equation

$$\frac{d^\alpha}{dt^\alpha} n(t) + Cn(t) = 0. \quad (10)$$

Where $\frac{d^\alpha}{dx^\alpha}$ is the fractional Caputo derivative of order α defined as:

$$\frac{d^\alpha f}{dx^\alpha}(x) = \frac{1}{\Gamma(1-\alpha)} \int_0^x (x-\xi)^{-\alpha} \frac{df}{d\xi} d\xi, \qquad 0 < \alpha \le 1.$$

Equation (10) is known to have analytic solution in term of Mittag-Leffler function [17]. In which follows, we will use solutions of the model Eqs. (6) and (10).

5. First result

5.1 Curve fitting

In the Fig. 6, we present the result of fitting procedure of the data experiment with the two relaxation models as presented in former section. The fitting method was carried out by optimization of a cost function measuring the distance between experimental data and models defined as:

$$F = ||n - Q_{FG}||_2. \quad (11)$$

The Levenberg-Marquard algorithm is used in order to produce the parameter C for the exponential model and the parameters (α,C) for the fractional relaxation. In this case, two fits results are proposed. The first corresponds to the upper part of the data experiment decreasing while the second fits the lower part of the data corresponding to long time behavior.

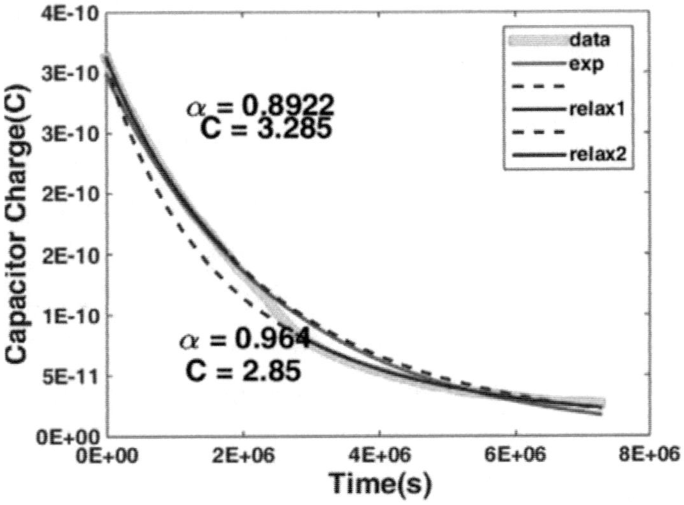

Figure 9. Different Relaxation models of charge loss (Dashed line indicates part of the M.F. approaches which does not fit the data dynamic).

5.2 Comments and discussions

Firstly, we note that the regime change observed in the section appears as the threshold region of the two schemes of fractional relaxation. We remark also that the exponential model used to fit the whole range of data does not match with this latter around this region. In the other hand, similar behavior is found for both fractional relaxation laws. These results suggest that two separate schemes may exist at two different time scales. Such conclusion has to be confirmed by the analysis of other datasets. If the result holds, we believe that an unified model is essential to depict the dynamic evolution of the charge loss over time. This may be reached by use of more complex random physical models such as anomalous diffusion process which takes account of possible long time trapping of particles inside the medium. Among, the possible candidates, we think that fractional Mobile Immobile model could be a suitable candidate.

In this case the main idea is to enforce a stochastic model that uses few parameters to describe charge transit through the oxide as a random walk. This approach to charge transport tentatively allows, from a macroscopic point of view, to link the observed evolution laws to dispersive (and possibly fractional) transport models [13]. This kind of models has been successfully used for mass transit through complex porous systems [14] as wells as other diffusive phenomena [15].

From electrical viewpoint, two complementary physical theories are known to describe these two schemes of current leakage behavior, namely the Fowler Nordheim (F-N) and the Poole Frankel (P-F) currents. The F-N current corresponds to a leakage process that appears at large value of floating gate voltage (VFG>3V) while the P-F current known also as the trap assisted tunneling current is predominant at low voltage range. As said in introduction, owing to the fact that these electrical models cannot be reached at very low current values, a valuable task for the future is to link our stochastic modeling that is easily available from datasets, with these electrical models (F-N and P-F).

6. Conclusion

Using classical protocol of the Floating Gate measurement, we are developing a new model of charge loss. Works still in progress. The main interest of our model comes from its robustness requiring only few parameters. In the other hand, electrical models of P-F and F-N need efficient leakage current measurement that cannot be reached from numerical derivation of charge loss data. In this context, our stochastic model allows one to derive the charge loss dynamic in a statistical sense.

References

[1] W. Brown and J. Brewer, "Nonvolatile semiconductor memory technology: a comprehensive guide to understanding and to using NVSM devices," IEEE Press, 1998.
[2] P. Canet, R. Bouchakour, N. Harabech, P. Boivin and J.-M. Mirabel, "Eeprom programming study-time and degradation aspects," Proc. of IEEE ISCAS, pp. 846–849, 2001.
[3] R. Laffont, R. Bouchakour, O. Pizzuto and J.-M. Mirabel, "A 0.18 um flash source side erasing improvement," Proc. of IEEE NVMTS, pp. 105–109, 2004.
[4] P. Canet, R. Bouchakour, N. Harabech, P. Boivin, J.-M. Mirabel and C. Plossu, "Study of signal programming to improve eeprom cells reliability," Proc. of IEEE MWSCAS, pp. 1144–1147, 2000.
[5] P. Canet, R. Bouchakour, J. Razafindramora, F. Lalande and J.-M. Mirabel, "Very fast eeprom erasing study," Proc. of IEEE ESSCIRC, pp. 683–686, 2002.
[6] ITRS, "Process Integration, Devices, and Structures", 2013.
[7] B. Fishbein, D. Krakauer and B. Doyle, "Measurement of Very Low Tunneling Current Density in SiO, Using the Floating- Gate Technique," IEEE Elec. Dev. Lett., **12**, no. 12, pp. 713–715, 1991.
[8] F.H. Gaensslen and J.M. Aitken, "Sensitive Technique for Mea- suring Small MOS Gate Currents," IEEE Elec. Dev. Lett., **1**, no. 11, pp. 231–233, 1980.
[9] Keysight B1500A Semiconductor Device Analyzer data sheet, October 2016.
[10] S. Burignat, "Mécanismes de transport, courants de fuite ultra-faibles et rétention dans les mémoires non volatiles à grille flottante," PhD thesis, INSA Lyon (France), 2004.
[11] J. Postel-Pellerin, G. Micolau, P. Chiquet, J. Melkonian, G. Just, D. Boyer and C. Ginoux, "Setting up of a floating gate test bench in a low noise environment to measure very low tunneling currents," Acta Imeko, **4**, no. 3, pp. 36–41, 2015.
[12] J. Postel-Pellerin, G. Micolau, C. Abbas, P. Chiquet and A. Cavaillou, "Robustness of the floating-gate technique in a very low-noise environment," Proc. of IEEE CAS, pp. 287–290, 2014.
[13] M.-C. Néel, A. Zoia, M. Joelson, A. Cartalade and M. Fleury, "Fractional p.d.e and stochastic processes for dispersion: application to porous media," Proc. of "Dynamiques fractionnaires et Applications" Conference, 2010.
[14] J. Golder, M. Joelson and M.-C. Néel, "Mass transport with sorption in porous media," Mathematics and Computers in Simulation, **81**, pp. 2181–2189, 2011.
[15] J. Golder, M. Joelson, M.-C. Néel and L. Di Pietro, "A time fractional model to represent rainfall process," Water Sci. and Engineering, **7**, no. 1, pp.32–40, 2014.
[16] N. Laskin, "Fractional Poisson Process" Communication in Nonlinear Science and Numerical Simulation, **8**, pp. 201–213, 2003.
[17] D. Sierociuk, I. Podlubny and I. Petras "Experimental Evidence of Variable-Order Behavior of Ladders and Nested Ladders". IEEE Transactions on Control Systems Technology **21**(2), 2011.
[18] Kilbas AA, Saigo M, "Generalized Mittag-Leffler function and generalized fractional calculus operators". Integral Transform Spec. Func. **15**(1):31–49, 2004.

Detection of alpha particle contamination on ultra low activity-grade integrated circuits

Ana C. Fernandes[1,a], Tomoko A. Morlat[1], Miguel Felizardo[1], Andreas Kling[1], Raul C. Martins[2], José G. Marques[1], Ana R. Ramos[1], Ignácio Lázaro[3], Thomas A. Girard[4], and Austin Lesea[5]

[1] Centro de Ciências e Tecnologias Nucleares, Instituto Superior Técnico, Universidade de Lisboa, 2695-066 Bobadela, Portugal
[2] Instituto de Telecomunicações, Instituto Superior Técnico, Universidade de Lisboa, 1049-001 Lisboa, Portugal
[3] Laboratoire Souterrain à Bas Bruit, University of Nice, University of Avignon, Centre National de la Reserche Scientifique, Aix-Marseille University, Observatoire de la Côte d'Azur, 84400 Rustrel, France
[4] Departamento de Física, Faculdade de Ciências, Universidade de Lisboa, 1749-016 Lisboa, Portugal
[5] Xilinx Inc., San Jose, CA 95124, USA

Abstract. We propose to apply the superheated droplet detector (SDD) technology to the measurement of alpha-particle emissivity on integrated circuits of ultra-low activity grade ($< 1\alpha$/khcm^2) for high reliability applications. This work is based on the SDDs employed within our team to the direct search for dark matter. We describe the modifications in the dark matter SDDs with respect to fabrication, signal analysis and characterization, in order to obtain a device with the adequate detection sensitivity and background noise.

1. Introduction

Soft Errors (SE) are non-destructive functional errors induced by energetic ion strikes. An important parameter is the critical charge required to change the logical state of a stored data bit. With the continuous technology downscaling, the decreased node capacitances and operating voltages reduce the critical charge, thereby increasing device sensitivity to SE. Electronic devices in aircrafts, space missions and nuclear power plants are specially selected for their reduced intrinsic SE rate (SER) and resistance against radiation-induced SE.

There are two primary sources of terrestrial SE: alpha-particles (α) emitted from the radioactive impurities in materials nearby the sensitive volume (packaging, solder bumps, etc) and atmospheric neutrons (n) which produce highly ionising charged particles on their interaction with the component materials. The α-induced SER increases more rapidly with decreasing critical charge and can overcome the n-induced SER even with Ultra Low Activity (ULA) materials, i.e., with alpha emissivity below $10^{-3}\alpha\ h^{-1}cm^{-2}$ [1].

Although fast and thermal neutron-induced SE reappeared in advanced integrated circuits (IC) [1, 2] they can be kept negligible even below 28 nm with improved Si technology and carrier gases depleted in ^{10}B [3]. Alpha contamination remains therefore a major concern of nanoelectronic companies for commercial, industrial, aerospace and defense applications

[a] Corresponding author: anafer@ctn.tecnico.ulisboa.pt

where reliability is mandatory. Qualification methodologies for ULA ICs have included variable-dependent measurements or underground tests (in a place devoid of cosmic radiation with negligible local α- and n-backgrounds). Evidence of any contamination is apparent in less than 6 months underground, but measuring the α-SER with any precision after 1–2 year testing is not possible with any of these methods [4].

Industry roadmaps call for instruments with detection limits of 10^{-3} α $h^{-1}cm^{-2}$ in the energy range 1–10 MeV for measurement times of less than 1 week with sample sizes $\sim$1500 cm^2 at a cost < 50 kEUR [5]. These conditions cannot be fulfilled by any of the current commercial proportional counters, since their backgrounds (5 count/h) are a factor of 5 too high [6]. Promising alternatives based on ionisation counters are under development and evaluation for ultra-low background α-emissivity measurements, but no general consensus has been obtained up to now. One of the problems is the internal contamination by radon-emitting isotopes that simulate α's from the sample and moreover the cosmic bias that induces a lab-lab variability of the measurement accuracy [7, 8].

In this work we propose a new approach to measure α-emission rates, based on the superheated droplet detectors (SDD) developed by our team towards the direct search for dark matter (DM) [9]. Similarly to IC, SDD are sensitive to α's and n's. Until now, their development aimed to achieve minimal n-background signal, as events caused by DM mimic n-interactions and are extremely rare ($\ll$ 1 count/y in 1 kg target material). In the proposed application the focus is reversed, having α-detection as its main purpose. As our current SDDs can embed small samples of a few cm^2, we will primarily consider the measurement of individual devices. A target value of $< 10^{-3}\alpha\ h^{-1}$ for the technique sensitivity is thereby set.

Underground testing of the standard DM SDD demonstrated an intrinsic background in the order of 10^{-4} count/h and 10^{-3} count/h for n and α, respectively [9] with a detection efficiency of $\sim$95% [10]. The background signal being the critical parameter to qualify an instrument's ability for low-level measurements, our SDD has an intrinsic α-background signal 100 times smaller than that of the ionisation counter, and a comparable detection sensitivity.

In this paper we describe the main challenges towards the implementation of the new methodology and report preliminary results.

2. Ongoing research and development

2.1 Detector fabrication

Intensive R&D is underway towards the elaboration of an SDD with a sample embedded in a central position. The close contact between the droplets and the sample is mandatory, as α attenuation by any interface material results in a significant sensitivity loss. On the other hand the introduction of irregularities in a moderately superheated material triggers spontaneous nucleations that produce acoustic noise and reduce detector lifetime. We have recently defined a post-fabrication method based on a liquid detector material and recompression stages that yield a smooth sample-detector interface with a distribution of droplets in the vicinity of the sample. Operation at a lower superheat [11] via modification of the thermodynamic parameters or the superheated liquid is under consideration aiming at further noise reduction.

2.2 Event localisation and discrimination

Alphas emitted by the sample are located in its vicinity, in the central volume of the detector. In contrast, those originating in Rn diffused through the detector walls and from emitters

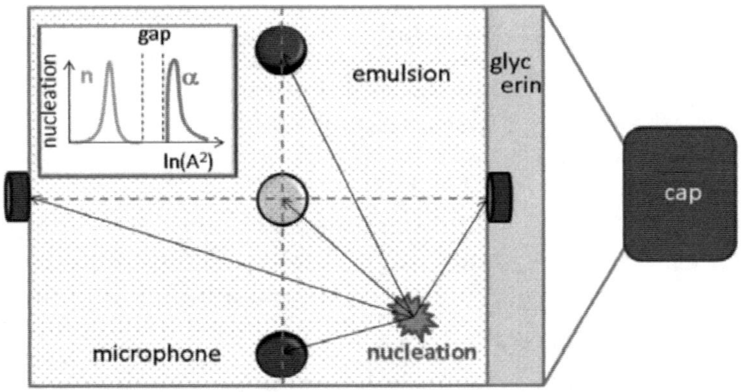

Figure 1. Schematic view of the SDD (in horizontal position) with a 5-microphone arrangement for the spatial localisation of the nucleations. The inset graph represents the amplitude (A) histograms of the acoustic signals corresponding to nucleations induced by recoiling nuclei following neutron interactions (n) and alpha-induced events (α), clearly separated by a gap.

in the detector container are localised near the glass-gel interface [12]. A technique towards the spatial localisation of events was developed based on the analysis of time delays among the signal detection by the various microphones of an array. Tests with a hot probe yielded a spatial resolution of 2 cm^3 that is adequate to identify events originating from a fiducial volume surrounding the sample [13].

Furthermore, the evaluation of the acoustic signals revealed that α- and n-events produce different amplitudes, larger ones corresponding to α's as a result of a greater bubble expansion power from the formation of various proto-bubbles. At signal amplitudes of ~100 mV there is a 20 mV gap between the α and n- distributions which is clearly resolved by the signal acquisition system (resolution 0.3 mV) [9].

We will proceed with the concomitant application of the localisation and particle discrimination routines in order to extract a signal free from any radiation rather than α's from the volume of interest, thereby decreasing noise and increasing device α-sensitivity. A schematic view of the detection/discrimination system is given in Fig. 1. Current investigations are focussed on the fine acoustic characterisation of the detector (e.g. model of sound propagation in the detector material, reflections in the detector container, dependence of sound speed with thermodynamic conditions) for an improved event localisation.

2.3 Response to alpha particles

Most SDD applications are concerned with the detection of neutrons or heavy recoiling atoms [11], whereas in this work we focus on the detection of α's. Our initial investigations of the intrinsic SDD response to α's were performed using spectroscopy-grade liquid sources that are homogeneously dispersed in the gel during the stirring phase of the fabrication process. Different α emitters (Sm, U and Th) and activities are considered. The nucleation rate as a function of temperature for $^{238/232}U$ in a C_2ClF_5 SDD at 1 bar is shown in Fig. 2, which displays a different variation from that corresponding to the local radiation background (measured with an undoped SDD).

Different bubble shapes and sizes and moreover different detection efficiencies were observed for the various emitters. The influence of α energy on the detector response is under investigation using a simplified model of the SDD response that considers the α Bragg

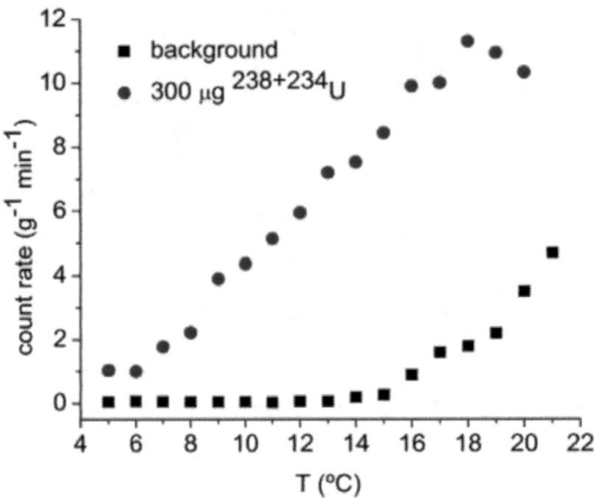

Figure 2. Temperature dependence of the response of a 150 ml SIMPLE SDD, doped with a 1mg/ml U solution and undoped (background).

peak [14] and will be further developed via Monte Carlo simulations coupled to stopping power and energy deposition calculations [15]. The correlation between bubble geometries and the acoustic signals will be evaluated. Such experiments using liquid sources will be explored to extract important parameters (e.g. Harper parameter [16]) for modelling detector response.

2.4 Neutron-induced alpha background

For SDD measurements at ground-level, production of α's within the fiducial sample volume upon neutron interaction with light elements should be addressed. The detector gel was evaluated by Ion Coupled Plasma Mass Spectroscopy with respect to B and Li and for completion, Sm, U and Th. Only upper limits were obtained corresponding to the measurement blanks: 5 ppm Li, 30 ppm B, 1 ppm Sm, 1 ppm U and 10 ppm Th. The upper limits for α emitters correspond to an intrinsic α-background much larger than measured. As for B, the measured upper limit corresponds to a production rate of alphas in the fiducial volume of a $(1.5 \times 1.5)\,\mathrm{cm}^2$ sample, exposed to $\sim$1 $n_{th}\,\mathrm{cm}^{-2}\mathrm{h}^{-1}$ thermal neutrons at ground-level [17] of $\sim 10^{-2}\alpha\,\mathrm{h}^{-1}$, which exceeds the order of magnitude of the target values for ULA levels. Measurements with improved detection sensitivity are therefore necessary. Furthermore, the variability of neutron fluence rates (geomagnetic, terrain, altitude, solar, weather) justify an evaluation of the neutron environment at the detector location. Although we favour the application of standard techniques such as Bonner spheres, the SDD themselves can provide an alternative methodology for environmental neutron spectrometry [18].

3. Future developments

Additional questions to be addressed include the investigation of the SDD response to solid α sources, and a series of test measurements at ground-level and underground – the latter intended to qualify the developed methodology. Measurements will initially focus on standard, calibrated α sources and gradually include actual integrated circuits. We are primarily focused on evaluating the following Field Programmable Gate Arrays from

Xilinx [19]: (i) Spartan 6 (45 nm), which has been tested underground with published α-SER values on the order of 135 FIT (1 FIT=1 error per 1E9 hours of use); (ii) Virtex 6 (40 nm), with a particularly low α-SER of 9 FIT/Mb; (iii) 7-series parts (28 nm) with α-SER=22 FIT/Mb; (iv) 20 nm and possibly 16 nm devices (Virtex and Kintex UltraScale). necessary.

Ideally, the underground measurements will be performed at the site where Phase II of SIMPLE was run (GESA room of Laboratoire Souterrain a Bas Bruit) in order to take advantage of the work made in the past towards the reduction and characterisation of the environmental neutron and alpha radiation levels. An effort was made to suppress Rn in the detector by continuous purging of the room air, circulation of the water bath and detector pressurization. The alpha background achieved (measured: $2 \times 10^{-3} \alpha$ h^{-1}) has order of magnitude of the aimed sensitivity [9]. As the value agrees with estimates of the Rn contribution, we expect to suppress the majority of the background α-events using the spatial localisation algorithms. The environmental neutron field in GESA (originating mainly from radioimpurities in its concrete walls) was reduced by two orders of magnitude with the installation of a water-based shielding [20]. The residual neutron background was characterised in detail using Monte Carlo simulations, their outputs being normalized to the measured activities of U/Th in the experiment materials [21]. The on-detector thermal neutron fluence rate of $\sim 1 \times 10^{-4}$ n_{th} cm^{-2} h^{-1} (calculated) will induce a negligible rate of α-events in the sample region (see Section 2.4). The fast neutron background (calculated: 2×10^{-4} n cm^{-2} h^{-1}) induces a reduced event rate in the detector (measured: 9×10^{-5} evt h^{-1}) that does not compromise its lifetime and can be rejected based on signal analysis.

4. Conclusions

We propose an innovative approach to the measurement of α-emissivity on ULA-grade integrated circuits, using the SDDs developed by our team for dark matter search. SDDs are of interest because its measured intrinsic α-background signal is two orders of magnitude lower than that of the most promising device under development for industry application, with a competitive sensitivity.

Additional noise reduction is achieved through the simultaneous application of spatial localisation and particle discrimination routines. The anticipated spatial resolution allows the measurement of individual (e.g. discard) devices rather than the large finished wafers evaluated by counters.

The fabrication of SDDs embedding a sample at both reduced spontaneous nucleation rate and high sensitivity is under investigation. Modifications relatively to the standard SIMPLE fabrication protocol are necessary, requiring a revision of the SDD response parameters. The application of liquid sources homogeneously distributed in the detector volume provides a benchmark for the investigation and verification of the physics underlying the detector operation.

In order to evaluate SDD noise at ground-level, the gel contamination by elements with significant (n, α) cross sections must be quantified with detection limits improved relatively to the measurements herein reported. The noise evaluation may benefit from the characterisation of the neutron environment at the SDD location.

In the final phase of the project, samples will be measured in an underground facility for the qualification of the methodology. The neutron background in the GESA room of LSBB has been characterised and is known to be, together with that from α's, appropriately low to conduct the experiment.

The authors acknowledge Fundação para a Ciência e Tecnologia (Portuguese Ministry of Science) for the support via grants IF/00628/2012/CP0171/CT0008 and SFRH/BPD/94028/2013, and research projects UID/Multi/04349/2013 and PTDC/EEI-ELC/2468/2014.

References

[1] H. Kobayashi, N. Kawamoto, J. Kase, et al., Proc. 2009 IEEE Intl. Reliability Physics Symp., 206 (2009).
[2] S.J. Wen, R. Wong, M. Romain, et al., Proc. 2010 IEEE Intl. Reliability Physics Symp., 1036 (2010).
[3] Y.S. Fan, A.S. Oates, IEEE Trans. Device Mat. Rel., **14**, 583 (2014).
[4] A. Lesea, K. Castellani-Coulié, G. Waysand, et al., IEEE Trans. Nucl. Sci., **55**, 2148 (2008).
[5] International SEMATECH Technology Transfer #01054118A-XFR (2001); #030 24377A-TR (2003).
[6] B. Clark, Proc. 1st IEEE-SCV Soft Error Rate Workshop, 2009 (`http://www.cpmt.org/scv/meetings/cpmt0910s.html`).
[7] M.S. Gordon, D.F. Heidel, K.P. Rodbell, IEEE Trans. Nucl. Sci. **56**, 3381 (2009).
[8] B. Clark, S. Coleman, B. McNally, 6th IEEE SER Workshop, 2014 (`http://www.cpmt.org/scv/meetings/cpmt1410s.html`).
[9] M. Felizardo, T.A. Girard, T. Morlat, et al., Phys. Rev. D, **89**, 072013 (2014).
[10] J.I. Collar, J. Puibasset, T.A. Girard, et al., Phys. Rev. Lett., **85**, 3083 (2000).
[11] F. d'Errico, Nucl. Instrum. Meth., **B184**, 229 (2001).
[12] A. Sonnenschein, 2009 Intl. Conf. Topics in Astroparticle and Underground Physics, 2009 (`http://taup2009.lngs.infn.it/slides/jul2/sonnenschein.pdf`).
[13] M. Felizardo, R.C. Martins, T.A. Girard, Comput. Stand. Inter., **32**, 96 (2010).
[14] M. Felizardo, T.A. Girard, A. Kling, et al., Astrop. Phys. (to be published).
[15] S. Seth, M. Das, J. Instrumentation, **11**, P04015 (2016).
[16] S. Archambault, F. Aubin, M. Auger, et al., New J. Phys., **13**, 043006 (2011).
[17] J.D. Dirk, M.E. Nelson, J.F. Ziegler, et al., IEEE Trans. Nucl. Sci., **50**, 2060 (2003).
[18] A.R. Ramos, F. Giuliani, M. Felizardo, et al., Radiat. Prot. Dosim., **115**, 398 (2005).
[19] Device Reliability Report – 2nd half 2015, Xilinx (2016) (`http://www.xilinx.com/support/documentation/user_guides/ug116.pdf`).
[20] A.C. Fernandes, M. Felizardo, A. Kling, et al., Nucl. Instrum. Meth. **A623**, 960 (2010).
[21] A.C. Fernandes, A. Kling, M. Felizardo, et al., Astrop. Phys., **76**, 48 (2016).

Seismic wavefield polarization – Part I: Describing an elliptical polarized motion, a review of motivations and methods

Claire Labonne[1,2,a], Olivier Sèbe[1], and Stéphane Gaffet[2,3]

[1] CEA, DAM, DIF, 91297 Arpajon, France
[2] Univ. Nice Sophia Antipolis, CNRS, IRD, Observatoire de la côte d'Azur, Géoazur UMR 7329, Valbonne, France
[3] LSBB UMS 3538, Rustrel, France

Abstract. The seismic wavefield can be approximated by a sum of elliptical polarized motions in 3D space, including the extreme linear and circular motions. Each elliptical motion need to be described: the characterization of the ellipse flattening, the orientation of the ellipse, circle or line in the 3D space, and the direction of rotation in case of non-purely linear motion. Numerous fields of study share the need of describing an elliptical motion. A review of advantages and drawbacks of each convention from electromagnetism, astrophysics and focal mechanism is done in order to thereafter define a set of parameters to fully characterize the seismic wavefield polarization.

1. Introduction

The seismic wavefield is a combination of polarized waves in the three-dimensional (3D) space. The polarization is a characteristic of the wave related to the particle motion. The displacement of particles effected by elastic waves shows a particular polarization shape and a preferred direction of polarization depending on the source properties (location and characteristics) and the Earth structure. P-waves, for example, generate linear particle motion in the direction of propagation; the polarization is thus called linear. Rayleigh waves, on the other hand, generate, at the surface of the Earth, retrograde elliptical particle motion. Three-component seismic stations record the trajectory of a particle displacement at the free surface of the Earth, and thus the polarization. The change in polarization can be analyzed and interpreted to improve our understanding of both the source process and the Earth structures illuminated by the recorded waves. The objective of the polarization analysis is the determination of the state of polarization consisting in a type of polarization (linear, purely elliptical or circular) and an orientation of the polarization in 3D space. This polarization analysis requires a system of parameters fit to describe an object and its attributes in space. Numerous fields of study share this need. In the following, in order to be able to build an efficient parameter system in seismology, a reviewed convention from other disciplines (electromagnetism, astrophysics and focal mechanism) is done and a parallel between these disciplines and seismology is drawn.

[a] Corresponding author: claire.labonne@cea.fr

2. Why an elliptical model?

A 3-component record of a monochromatic wave can be expressed as followed:

$$\begin{pmatrix} x \\ y \\ z \end{pmatrix}(t) = \begin{pmatrix} A_x\cos(2\pi ft - \varphi_x) \\ A_y\cos(2\pi ft - \varphi_y) \\ A_z\cos(2\pi ft - \varphi_z) \end{pmatrix} = \underbrace{\begin{pmatrix} A_x\cos\varphi_x \\ A_y\cos\varphi_y \\ A_z\cos\varphi_z \end{pmatrix}}_{\vec{p}} \cos(2\pi ft) + \underbrace{\begin{pmatrix} A_x\sin\varphi_x \\ A_y\sin\varphi_y \\ A_z\sin\varphi_z \end{pmatrix}}_{\vec{q}} sin(2\pi ft) \quad (1)$$

where A_i and φ_i represent the amplitude and the phase of the component i.

This expression corresponds to the equation of an ellipse as shown in [1]. Thus, the trajectory of a particle through the propagation of a monochromatic wave describes necessarily a full ellipse as a function of time every one time period T=1/f. The vectors $\vec{p}$ and $\vec{q}$ can be seen as the positions on the ellipse at t=0 and at t=T/4. The extreme cases of linear and circular polarizations will be described by respectively collinear $\vec{p}$ and $\vec{q}$ (linear case) and $|\vec{p}| = |\vec{q}$ (circular case).

Using Fourier analysis, an arbitrary real signal can be decomposed into a sum of quasi-monochromatic signal:

$$\begin{pmatrix} x \\ y \\ z \end{pmatrix}(t) = 2\int_0^{\infty}\int_{-\infty}^{+\infty} \vec{p}(f,\tau)\cos(2\pi ft) + \vec{q}(f,\tau)\,sin(2\pi ft)\,d\tau df \quad (2)$$

where $\vec{p}(f) = \begin{pmatrix} Re(X(f,\tau)) \\ Re(Y(f,\tau)) \\ Re(Z(f,\tau)) \end{pmatrix}$, $\vec{q}(f) = -\begin{pmatrix} Im(X(f,\tau)) \\ Im(Y(f,\tau)) \\ Im(Z(f,\tau)) \end{pmatrix}$, $X(f,\tau)$ corresponds to a local Fourier transform of x(t) e.g. the Stockwell transform [2], *Re* and *Im* correspond to real and imaginary part operator.

The seismic wavefield records can be decomposed into a sum of quasi-monochromatic waves, and thus a sum of elliptical motions. To describe each elliptical motion, the characterization of the ellipse flattening, the orientation of the ellipse, circle or line in the 3D space, and the direction of rotation must be defined.

3. Polarization and orientation conventions in other disciplines

3.1 Electromagnetism

In physics, the notion of polarized waves is not only defined in seismology. In electromagnetism, wave polarization is defined by the temporal behavior of a transverse electromagnetic wave at a given point in space. The state of polarization is investigated in the plane perpendicular to the wave's direction of propagation. Two parameters are introduced to characterize this elliptical state of polarization (see Fig. 1) [1]:

- the signed ellipticity ($\boldsymbol{\rho_S}$) or the ellipticity angle ($\boldsymbol{\chi} = \text{atan}(\rho)$)
  This parameter specifies the shape and the direction of rotation. It is defined as the ratio between the semi-minor axis (***b***) and the semi-major axis (***a***): $\rho_S = \pm\frac{b}{a}$. The signed ellipticity values (respectively ellipticity angle) range from –1 to 1 (resp. from $-\frac{\pi}{4}$ to $\frac{\pi}{4}$). A zero value (resp. 0) corresponds to a purely linear motion, and ± 1(resp. $\pm\frac{\pi}{4}$) to purely circular motion. The sign of ρ_S (resp. χ) discriminates the two possible directions in which the ellipse may be described: handed-left or handed-right polarization.
- the inclination $\boldsymbol{\gamma}$
  This parameter represents the angle between the major direction and the reference direction. The inclination values range from 0 to π.

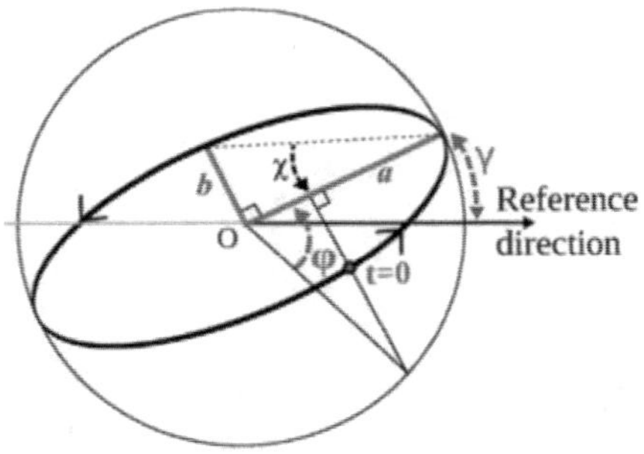

Figure 1. Electromagnetism parameters convention. The blue dot corresponds to the position on the ellipse at time t = 0.

Linear and circular polarizations are special cases of elliptical polarization. The "purely elliptical polarization" notion will be used to exclude these two special cases. Each individual state of polarization is defined by a unique pair of shape-orientation parameters $(\boldsymbol{\chi}, \boldsymbol{\gamma})$. A graphical representation can be done using Poincaré's sphere on which all states of polarization are compressed over the surface of a sphere [3].

In order to reconstruct and not only characterize a polarized motion, one additional parameter needs to be defined, the phase factor $(\boldsymbol{\varphi})$. The phase permits to locate the point on the ellipse at t=0. Consider a circumscribed circle with a radius of a, that is concentric with an ellipse of semi-major axis of a. A line perpendicular to the major axis of the ellipse is extended through the point at t=0 and intersects the circle. The phase factor is the angle between the intersect point and the major direction, thus, the eccentric angle and not the polar angle of the position at t=0 relatively to the major direction.

To parametrize an elliptical motion in 2D plane, four parameters are required: the semi-major axis a, the signed ellipticity ρ_s, the inclination γ and the phase factor φ:

$$\begin{bmatrix} x \\ y \end{bmatrix}(t) = \begin{bmatrix} \cos\gamma & -\sin\gamma \\ \sin\gamma & \cos\gamma \end{bmatrix} \begin{bmatrix} a\,\cos(2\pi f t - \varphi) \\ \underbrace{a\rho_s}_{\pm b}\,\sin(2\pi f t - \varphi) \end{bmatrix} \tag{3}$$

The parameter system used to describe the polarization in electromagnetism fully characterizes the shape of polarization, the direction of rotation and the orientation of polarization relative to a reference plane. However, this system requires the a-priori knowledge of the orientation of the polarization plane which would be a major limitation in seismology. In seismology, the seismic wavefield is a combination of polarized waves in the 3D space. Due to multi-pathing propagation, a plane of reference cannot be pre-defined. Therefore, polarization parameters need to be oriented in the 3D space without pre-defined plane of polarization.

3.2 Astrodynamics

In a gravitational two-body problem, the relative position of one body with respect to the other follows an elliptical orbit in 3D space [4]. Therefore, the particle motion in seismology can be considered analogous to the elliptical orbit of bodies in space. The relative orbit motion is confined in a plane, but motions in solar system are not confined to a single plane. That is why, in astrodynamics, a 3D representation of an orbit in space is considered (see Fig. 2).

The orbital motion is referred to a reference plane in which the direction of a reference line is defined and called **x-direction**. The origin of the coordinate system is centered on one

DOI: 10.1051/e3sconf/20161206001

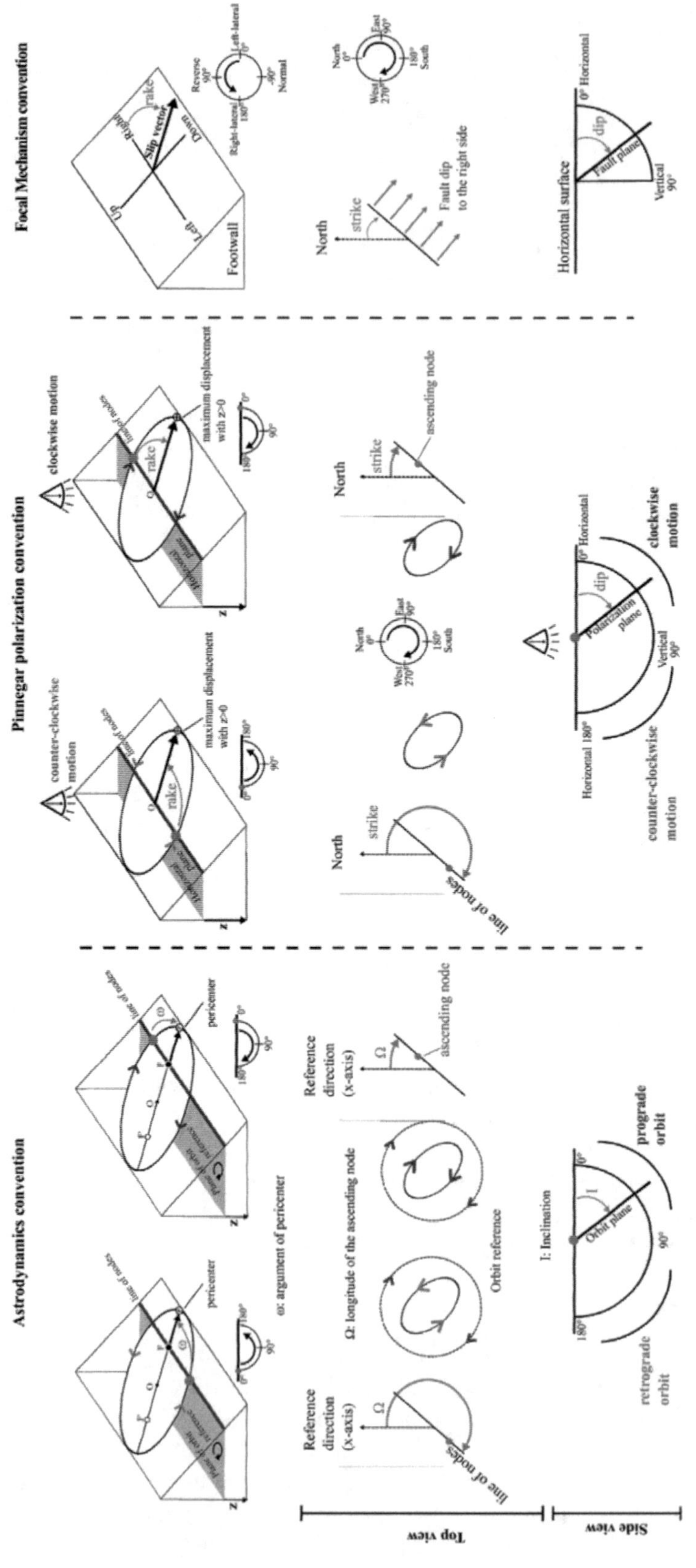

Figure 2. Comparison of the astrodynamics convention, the polarization parameter system of [5], and the focal mechanism convention.

of the two bodies at a focus of the ellipse. Usually, the reference plane is defined as the orbit plane of a reference object. For example, when considering the planets' orbit around the sun, a sun-centered coordinate system is chosen, the reference plane is defined as the Earth's orbit and the reference line is in the direction of the vernal equinox along the line of intersection of the Earth's equator and the Earth's orbit plane. This system can be used to describe any planet's orbit around the sun.

The **z-direction** is a vector orthogonal to the reference plane, which direction is defined by the rotation of the reference object. In the example mentioned above, the direction of rotation of the Earth is the defining criterion for the positive z-direction.

In general, orbit planes do not coincide with the reference plane but have an inclination angle called **inclination of the orbit, I**. The line of intersection between an orbital plane and the reference plane is called **line of nodes**. The point where an orbit crosses the reference plane with positive z-direction is called **ascending node**. The angle between the reference line and the radius vector of an ascending node is called **longitude of the ascending node, Ω**. The point corresponding to the minimum orbital radius is called the **pericenter**. The angle between the radius vector of the ascending node and the pericenter radius is called **the longitude of the pericenter, ω**.

An orbital plane can be described by a local right-handed triad system (i,j,k). The i-axis is taken to lie along the major axis in the direction of the pericenter. The j-axis is perpendicular to the i-axis and is confined in the orbital plane; its direction is defined by the direction of the rotation. The k-axis is taken perpendicular to the i- and j-axis.

To summarize, three angles (inclination I – argument of pericenter ω – longitude of ascending node Ω) are used to relate the local orbital plane system (i,j,k) with a general reference system (x,y,z) by a series of three rotations: a rotation around the z axis with an angle of ω, so that the x-axis coincide with the line of nodes, a rotation around the x-axis with an angle of I, so that the two planes are coincident, and finally a rotation around the z-axis with an angle of Ω.

Note that the inclination ranges from $0°$ to $180°$. If the inclination is smaller than $90°$ the motion is prograde whereas if the inclination is larger than $90°$ the motion is retrograde. A "purely" elliptical motion in the 3D space can be fully orientated using this coordinate system.

In seismology, the most complete polarization parametrization uses similar angular convention [5]. The only difference lies in the choice of the origin of the coordinate system defined at the center of the ellipse and not at a focus of the ellipse (Fig. 2). The reference plane is the horizontal plane and the direction of reference is the North (x-axis). The y-axis corresponds to the East and the z-vector is oriented downward. Since the origin of the coordinate system is now at the center of the ellipse, a pericenter cannot be defined anymore. In order to be able to orientate the ellipse, the notion of pericenter is replaced by the notion of "positive" maximum displacement which corresponds to the position of maximum displacement in the ellipse with a strictly positive z-coordinate. As in astrodynamics, three angles are defined to relate the local polarization plane system (i,j,k) with a general reference system (North, East, Down): **the rake, the dip and the strike**.

This system still has limitations. A purely horizontal polarization plane is a singularity which cannot be described as both maximum displacements would have the same z-coordinate making them impossible to differentiate. This parameter system requires the definition of both a plane of polarization and a major displacement.

In the case of linear motion, the motion is confined to a line meaning that neither the line of nodes nor the plane of polarization can be defined. Without a line of nodes none of the above mentioned angles exist and a new system would have to be used like the trend and plunge system in structural geology.

In the case of circular motion, the norm of the displacement vector is constant over the time, thus the position of major displacement cannot be defined and the rake becomes irrelevant. It is worth noting that the rake, the strike and the dip defined above do not exactly correspond to the rake, the strike and the dip used to orientate fault planes in focal mechanism and structural.

3.3 Focal mechanism

Fault strike, fault dip, and slip rake are used to describe earthquake focal mechanisms, consisting in fault plane orientation, and the direction of the slip displacement. The **fault strike** is the azimuth of a strike line, which is the line created by the intersection of a fault plane and a horizontal plane, relative to North. Strike is always defined such that a fault dips to the right side of the trace when moving along the trace in the strike direction. Fault strike values range from $0°$ to $360°$. The **fault dip** is the angle between the fault plane and a horizontal plane, with values ranging from $0°$ to only $90°$. The **slip vector** shows displacement of the hanging wall relative to the footwall, where the hanging wall is the fault block above the fault plane and the footwall the block below the fault plane. The **slip rake** is the angle between the strike direction and the slip vector. The rake values ranges from $-180°$ to $180°$.

Using this system for polarization characterization, the polarization plane corresponds to the fault plane, and the maximum displacement corresponds to the slip displacement. As defined the strike and the dip permit to orientate the plane in space. However, these two angles are not sufficient to characterize the direction of a rotating motion inscribed in the plane and thus orientate the normal vector to this plane. The direction of a rotating motion is an important parameter to characterize the polarization. To give an example, Rayleigh waves usually generate at the surface of the Earth retrograde elliptical particle motion, but the motion recorded at the surface can still sometimes be prograde, as a result of a specific Earth structure below the recorded station.

The major limitation of the focal mechanism system is its inability to characterize a direction of motion contrary to the astrodynamics convention. A comparison of the strike, dip and rake for both conventions is available in Fig. 2.

4. Conclusion

The seismic wavefield can be approximated by a sum of elliptical polarized motions in 3D space, including the extreme linear and circular motions. The study of the polarization is a key element in the analysis of source properties and Earth structure. For this study, a system of parameter needs to be defined to characterize the ellipse flattening, the orientation of the ellipse, circle or line in the 3D space, and the direction of rotation in case of non-purely linear motion. In electromagnetism, the waves' characteristics to be described are very similar to seismic waves'. Therefore, the parameter system would be very well adapted to our needs if it weren't limited by the 2D space hypothesis. To overcome the 2D limitation, the astrodynamics provides a full 3D orientation system, presenting the only drawback of having the origin located at a focus of the ellipse instead of the center. [5] proposed a system which regroups the advantage of both electromagnetism and astrodynamics conventions. Finally, [5] system must not be mistaken for the convention used in focal mechanism description. A standardized full parameter system is required in seismology (as defined in [6]) to provide a full characterization of any elliptical motion in 3D space including the extreme linear and circular motions and to provide access to all parameters required in most polarization studies.

References

[1] Born, M., Wolf, E. (2000), Principles of optics: electromagnetic theory of propagation, interference and diffraction of light. Cambridge University Press.

[2] Stockwell, R.G., Mansinha, L., & Lowe, R.P. (1996), Localization of the complex spectrum: the S transform, IEEE transactions on signal processing, **44**(3), 998–1001.

[3] Deschamps, G., & Mast, P. (1973), Poincaré sphere representation of partially polarized fields. IEEE Transactions on Antennas and Propagation, **21**(3), 474–478.

[4] Murray, C.D., & Dermott, S.F. (1999), Solar system dynamics, Cambridge University Press.

[5] Pinnegar, C.R. (2006), Polarization analysis and polarization filtering of three-component signals with the time—frequency S transform. Geophysical Journal International, **165**(2), 596–606.

[6] Labonne C., Sèbe O., Gaffet, S., Schindelé, F., Boyer D., Decitre J.B. and Cavaillou A., (2016) Seismic wavefield polarization – Part II: Definition of a parameter system in three-dimensional (3D) space, example case review using LSBB seismic station data.

 DOI: 10.1051/e3sconf/20161206002

Seismic wavefield polarization – Part II: Definition of a parameter system in three-dimensional (3D) space, example case review using LSBB seismic station data

Claire Labonne[1,2,a], Olivier Sèbe[1], Stéphane Gaffet[2,3], François Schindelé[1], Daniel Boyer[3], Jean-Baptiste Decitre[3], and Alain Cavaillou[3]

[1] CEA, DAM, DIF, 91297 Arpajon, France
[2] Univ. Nice Sophia Antipolis, CNRS, IRD, Observatoire de la côte d'Azur, Géoazur UMR 7329, Valbonne, France
[3] LSBB UMS 3538, Rustrel, France

Abstract. A full polarization parameter system in 3D space is presented to characterize the state of polarization of a seismic wavefield and to parametrize any type of elliptical polarized seismic wave including extreme linear and circular polarizations. This parameter system does not require the a-priori knowledge of the orientation of the polarization plane and provides access to all parameters required in most polarization studies. Two groups of angular and vectorial parameters are defined, which can be easily related to one another. The knowledge of particle motion polarization opens the field of applications for 3-component seismic records contributing to improvements in phase identification. An example of a time-frequency visualization of the polarization parameters on a teleseimic earthquake recorded at a seismic station of LSBB is presented.

1. Introduction

A 3-component seismogram corresponds to the local trajectory of a particle motion at the surface as recorded through the passing of the seismic wavefield. The particle motion trajectory varies with the type of wave. For example, the particle motion for a P-wave is linear along the direction of propagation and for an S-wave is orthogonal to the direction of propagation. The polarization is defined as a particle motion shape and orientation. The change in polarization can be analyzed and interpreted to improve our understanding of both the source process and the Earth structures illuminated by the recorded waves. The analysis of the polarization in seismology was traditionally performed by simple visual approach using hodochrones and displays of the 3-component seismogram. Still more quantitative approaches were developed e.g. [1–5]. However, each author used his own parameters and appellations depending on the type of wave to analyse. In this paper, we bring together and standardize the different polarization parameters in order to (i) characterize the state of polarization and (ii) model the particle motion for all types of polarization. The characterization and the modelling require a system of parameters fit to describe an object

[a] Corresponding author: `claire.labonne@cea.fr`

and its attributes in space. Numerous fields of study share this need. In the following, a full parameter system for seismology will be proposed based on the conclusions from [6] reviewing different parameter systems.

2. Seismic wave polarization parameter system

Starting the definition of the parameter system, the origin is defined at the center of the ellipse and the reference axis system is set to (*North, East, Down*).

To orientate an ellipse, a unique reference point needs to be chosen. Following [3], the positions of major displacement along the ellipse are ideal candidates. Still a criterion to discriminate between the two positions is needed since only one point can serve as reference. The **"positive" major displacement** defines the unique position of major displacement with a strictly positive Down-component i.e. the point is located under the reference plane. If the ellipse is parallel to the reference plane i.e. horizontal, this definition reaches a singularity which requires the introduction of additional criteria:

- In case of nullity of Down-component, the major displacement with strictly positive North-component is chosen.
- In case of nullity of both Down- and North-component, the major displacement with strictly positive East-component is chosen.

In case of purely circular motion, the notion of major displacement is irrelevant. This displacement has no orientation.

Deriving directly from the definition of the positive major displacement position, four parameters are defined:

- $\vec{a}$ vector: the "positive" major displacement position vector. $|\vec{a}|$, written as $\boldsymbol{a}$, is equal to the semi-major axis length. $\vec{a}$ corresponds to the position on the ellipse at the time t_a.
- $\vec{b}$ vector: the minor displacement vector with respect to the direction of rotation. $|\vec{b}|$, written as **b**, is equal to the semi-minor axis length. $\vec{b}$ corresponds to the position on the ellipse at t_a+ T/4, with T the rotation period.
- **Plunge $\boldsymbol{\Delta}$**: the angle from horizontal to the "positive" major displacement vector ($\vec{a}$).
- **Trend $\boldsymbol{\Omega}$**: the azimuth from North of the "positive" major displacement vector ($\vec{a}$).

Note that similar angular parameters to orientate the major displacement were also called dip or inclination and strike or azimuth by others authors as [1, 4, 5], not to be mistaken for the parameters defined in the next paragraph to orientate the polarization plane.

To orientate the ellipse plane and take into account the direction of rotation, an "oriented" ellipse plane shall be defined. Following the astrodynamics convention, the line of intersection between the ellipse plane and the horizontal plane is called the **line of nodes**. The **ascending node** is the point where a particle following the ellipse crosses the horizontal plane downward. In case of horizontal polarization plane, the notion of node is irrelevant. Any inclined plane in 3D space can be oriented using the following two angles:

- **Dip $\boldsymbol{\delta}$**: the angle between the horizontal plane and the ellipse plane.
- **Strike $\boldsymbol{\theta}$**: the azimuth from North of the ascending node.

A third independent angle, the **rake γ**, can be introduced to define directly the major direction relatively to the ellipse plane. It represents the angle between the ascending node and the

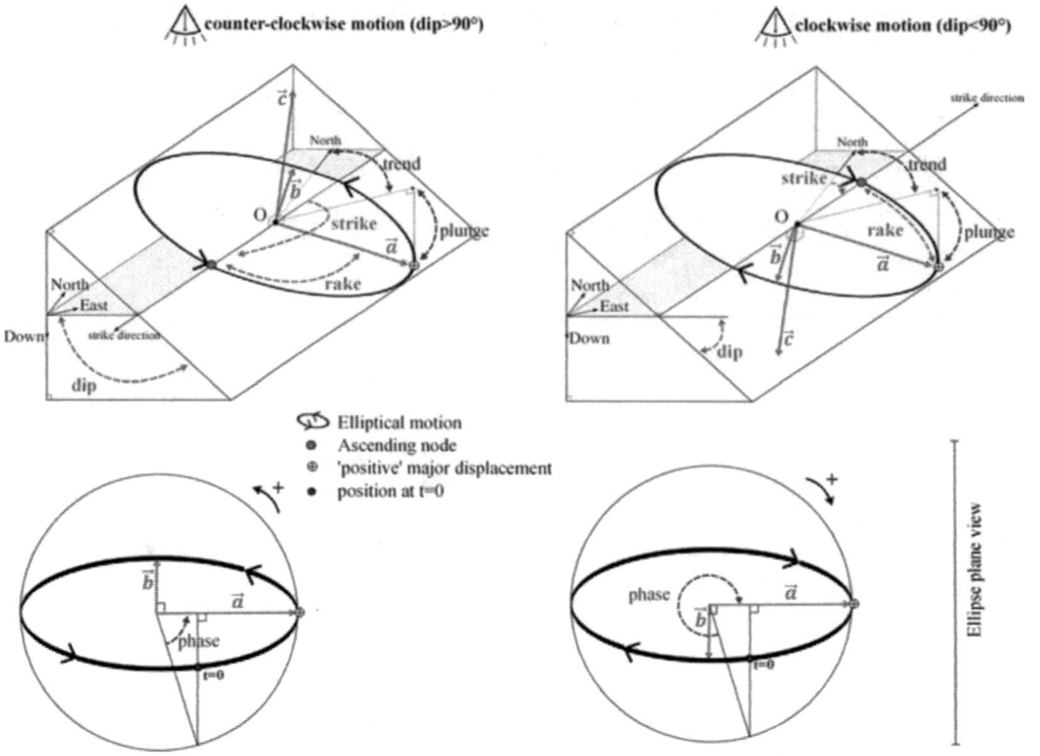

Figure 1. Seismic wave polarization parameter system.

"positive" major displacement. Note that these three angles depend on the direction of rotation. The triplets (δ, θ, γ) and $(\pi - \delta, \theta + \pi, \pi - \gamma)$ correspond to the same plane but define each a different direction of rotation (see Fig. 1).

To resolve the horizontal plane singularity, a vectorial approach is preferable to the angular approach above. The vectorial approach defines the following $\vec{c}$ vector, which is the normal vector to the ellipse plane taking into account the direction of rotation: $\vec{c} = \vec{a} \wedge \vec{b} = \vec{p} \wedge \vec{q}$. It presents the strong advantage of being a null vector in case of linear motion. The orientation of the polarization is fully controlled by the orientation of $\vec{a}$ and $\vec{c}$ vectors in case of purely elliptical motion, by $\vec{a}$ vector only for linear motion and by $\vec{c}$ vector only for circular motion.

As the angular and vectorial parameters are defined on the same basis, both group of parameters can easily be related to one another by the following equations:

$$trend = atan2(a_E, a_N) \tag{1}$$

$$plunge = asin\left(\frac{a_D}{|\vec{a}|}\right) \tag{2}$$

$$strike = atan2(c_N, -c_E) \tag{3}$$

$$dip = acos\left(\frac{c_D}{|\vec{c}|}\right) \tag{4}$$

where $\vec{a} = (a_N, a_E, a_D)$ and $\vec{c} = (c_N, c_E, c_D)$ in the (*North, East and Down*) system, atan2 is the four-quadrant inverse tangent.

 DOI: 10.1051/e3sconf/20161206002

Table 1. List of polarization parameters and their domain of validity.

PARAMETER	Particle motion		
	LINEAR	PURELY ELLIPTICAL	CIRCULAR
$\vec{a}$ vector	**unique**	**unique**	$(\vec{a}, \vec{b})$ non unique
$\vec{b}$ vector	$\vec{b} = \vec{0}$	unique	
$\vec{c}$ vector	$\vec{c} = \vec{0}$	**unique**	**unique**
Phase φ		$0 \leq \varphi < 2\pi$	φ depends on the choice of $\vec{a}$
a		$a > 0$	
b	$b = 0$	$0 < b < \text{a}$	$b = a$
Ellipticity $\rho = \frac{b}{a}$	$\boldsymbol{\rho = 0}$	$\boldsymbol{0 < \rho < 1}$	$\boldsymbol{\rho = 1}$
Dip δ	undefined	$\boldsymbol{0 \leq \delta \leq \pi}$	
Strike θ	undefined	$\boldsymbol{0 \leq \theta < 2\pi}$ undefined if $\boldsymbol{\delta} = 0$	
Rake γ	undefined	$0 \leq \gamma \leq \pi$ undefined if $\boldsymbol{\delta} = 0$	undefined
Plunge Δ		$\boldsymbol{0 \leq \Delta \leq \frac{\pi}{2}}$	undefined
Trend Ω		$\boldsymbol{0 \leq \Omega < 2\pi}$	undefined

Angular and vectorial approaches can both fully describe an ellipse in 3D Space. They, however, present different advantages and drawbacks. The vectorial approach presents the major advantage of removing the singularity from the angular approach in case of a horizontal polarization plane and can easily be used to recover the angular definition (see equations above). The angular approach is closer to the usual representation of space which makes it easier to visualize.

In addition to describing the polarization orientation, the second objective of the proposed set of parameters is to characterize the shape also called ellipticity. The ellipticity parameter (ρ) is defined as the ratio between b and a: $\rho = \frac{b}{a}$

To reconstruct and not only characterize a polarized motion, one additional parameter needs to be defined, the **phase factor** ($\boldsymbol{\varphi}$). As in electromagnetism, the phase permits to locate the point on the ellipse at t=0. The only difference is that the phase factor is an "orientated" angle relatively to the "orientated" polarization plane, meaning that the phase angle is in the same direction than the direction of rotation. Consider a circumscribed circle with a radius of a, that is concentric with an ellipse of semi-major axis a. A line perpendicular to the $\vec{a}$ vector is extended through the point at t=0 and intersect the circle. The phase factor is the angle between the intersect point and the $\vec{a}$ vector, also called the eccentric angle. The phase factor $\boldsymbol{\varphi}$ is the eccentric angle between the position at t=0 and the position of "positive" major displacement. This definition of the phase becomes irrelevant in case of circular motion because the notion of position of "positive" major displacement is irrelevant in such case.

All the parameters defined in this section are listed in the Table below, which summarizes the domain of validity of each parameter. The parameters that can be used to characterize the state of polarization (the shape or the orientation) are in bold.

To parametrize an elliptical motion in 3D space, the angular and vectorial approaches are equally valid and both require a set of six parameters. The six parameters proposed in case of the angular convention are the semi-major axis, the ellipticity, the dip, the strike, the rake and the phase factor, leading to the following parametrization [3]:

$$\begin{bmatrix} x_N \\ x_E \\ x_D \end{bmatrix}(t) = \begin{bmatrix} \cos\theta & -\sin\theta & 0 \\ \sin\theta & \cos\theta & 0 \\ 0 & 0 & 1 \end{bmatrix} \begin{bmatrix} 1 & 0 & 0 \\ 0 & \cos\delta & -\sin\delta \\ 0 & \sin\delta & \cos\delta \end{bmatrix} \begin{bmatrix} \cos\gamma & -\sin\gamma & 0 \\ \sin\gamma & \cos\gamma & 0 \\ 0 & 0 & 1 \end{bmatrix}$$
$$\times \begin{bmatrix} a\cos(2\pi ft - \varphi) \\ a\rho\sin(2\pi ft - \varphi) \\ 0 \end{bmatrix} \tag{5}$$

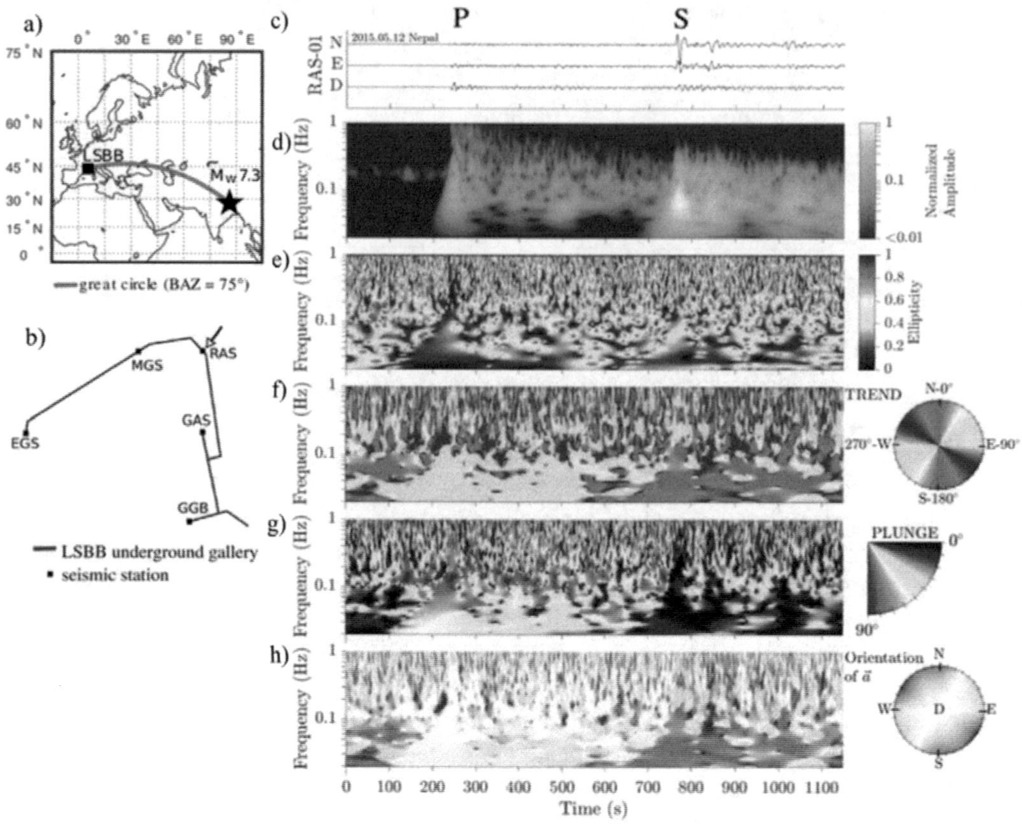

Figure 2. (a) Map showing the position of the earthquake relatively to the LSBB. (b) Map of LSBB seismic array, the arrow points the station used in this article named RAS (c) 3-component seismogram, N (North), E (East), D (Down) (d) Normalized amplitude spectrum obtained from an average of the 3-component Stockwell spectra (e) Ellipticity spectrum (f) Trend spectrum (g) Plunge spectrum (h) $\vec{a}$ vector orientation spectrum.

In case of the vectorial convention, the six parameters are the five independent components of the orthogonal vectors $\vec{a}$ and $\vec{c}$ and the phase factor, leading to the following parametrization:

$$\begin{bmatrix} x_N \\ x_E \\ x_D \end{bmatrix}(t) = \vec{a}\ \cos(2\pi ft - \varphi) + \underbrace{\vec{c} \wedge \vec{a}}_{\vec{b}}\ \sin(2\pi ft - \varphi) \quad (6)$$

This parametrization can be used to rebuild a polarized motion from information extracted from records.

3. Visualization and comparison of polarization parameters through an earthquake record at LSBB seismic station

To illustrate the parameter system defined above, a short study of the polarization of a seismic wavefield recorded on a Low Noise Underground Laboratory (LSBB, Laboratoire Souterrain à Bas Bruit) seismic station is presented in the following section. The example data correspond to the Mw 7.3 earthquake that occurred on May 12th 2015 in Nepal recorded by RAS station (see Fig. 2a and 2b). The epicentral distance and the back-azimuth associated to the recording station are respectively 77° and 75°. Figure 2c displays a zoomed-in section of the seismogram focusing on the first arrivals of P- and S- waves.

Following the time-frequency polarization analysis of [3], consisting in the Stockwell time-frequency decomposition (Fig. 2d) and in the approximation of each time-frequency contribution by an elliptical motion lying in a plane in the 3-D space, the polarization parameters are estimated directly from the 3-component data. Some of these parameters are represented in Fig. 2e-f-g-h, the ellipticity, the trend, the plunge, and $\vec{a}$ vector.

Assuming a 1D Earth model, the direct ray path from the epicentre to the recording station will have an azimuth of 75° and an incidence of 20°. For the direct P-wave, the major particle displacement direction is expected to be parallel to the ray path, corresponding to respectively a trend and the plunge of 77° and 70°. Low ellipticity values specific to linear motion (lower than 0.3), trend and plunge of approximatively 80° and 60–90° are observed at the direct P wave arrival time of 250 s roughly matching the expectations from the 1D Earth model.

For the direct S-wave, the major particle displacement direction is expected to be orthogonal to the ray path. In Fig. 2, the direct S-wave is observed at ~ 775 s on the seismogram with quasi-horizontal North-South particle motion, corresponding to an orthogonal direction to the expected ray path.

Figure 2h displays the orientation of the $\vec{a}$ vector. The hue corresponds to the trend and the saturation level to the plunge. This representation of $\vec{a}$ vector orientation regroups in only one figure the information of both trend and plunge. Similarly to $\vec{a}$ vector, a representation of the orientation of $\vec{c}$ vector can also regroup the information of dip and strike.

4. Conclusion

Polarization is an important property that can be used to better characterize both source and Earth structure from seismic recordings. Polarization can be used to further improve the extraction, the visualization, the filtering, and the rebuilding of the polarized seismic wavefield in 3D space. An exhaustive and simple parameter system is an absolute pre-requisite to the description of the polarization. Two equivalent groups of parameters have been defined, angular and vectorial parameters. The angular parameters can be deduced from the vectorial parameters. To fully characterize the polarization, both shape and an orientation should be described in space. The shape is fully controlled by the ellipticity parameter. Using the vectorial parameter system, the orientation of the polarization is fully controlled by the orientation of the $\vec{a}$ and $\vec{c}$ vectors in case of purely elliptical motion, by the $\vec{a}$ vector only for linear motion and by the $\vec{c}$ vector only for circular motion. The vectorial parameters have the strong advantage to reduce the number of parameter to visualize.

This work was performed through the funding of the "Région Provence Alples Côte d'Azur" (PACA) and the Commissariat à l'Energie Atomique et aux Energies Alternatives (CEA). Authors thank the LSBB team and the Technologie de la Mesure dans la Géosphère (CEA/TMG) staff for their technical help.

References

[1] Vidale, J.E. (1986), Complex polarization analysis of particle motion. Bulletin of the Seismological society of America, **76**(5), 1393–1405.

[2] Diallo, M.S., Kulesh, M., Holschneider, M., Scherbaum, F., & Adler, F. (2006), Characterization of polarization attributes of seismic waves using continuous wavelet transforms. Geophysics, **71**(3), V67–V77.

[3] Pinnegar, C.R. (2006), Polarization analysis and polarization filtering of three-component signals with the time—frequency S transform. Geophysical Journal International, **165**(2), 596–606.

DOI: 10.1051/e3sconf/20161206002

[4] Jurkevics, A. (1988), Polarization analysis of three-component array data. Bulletin of the Seismological Society of America, **78**(5), 1725–1743.

[5] Kulesh, M., Diallo, M.S., Holschneider, M., Kurennaya, K., Krüger, F., Ohrnberger, M., & Scherbaum, F. (2007), Polarization analysis in the wavelet domain based on the adaptive covariance method. Geophysical Journal International, **170**(2), 667–678.

[6] Labonne, C., Sèbe, O., Gaffet S. (2016), Seismic wavefield polarization – Part I: Describing an elliptical polarized motion, a review of motivations and methods.

EDP Sciences – Web of Conferences
17, Avenue du Hoggar
Parc d'Activité de Courtabouf
BP 112
F-91944 Les Ulis Cedex A
France

ISBN 978-1-5108-3361-6

CURRAN ASSOCIATES INC.
proceedings.com